AIR, WATER AND SOIL POLLUTION SCIENCE AND TECHNOLOGY SERIES

FROM SOIL CONTAMINATION TO LAND RESTORATION

AIR, WATER AND SOIL POLLUTION SCIENCE AND TECHNOLOGY SERIES

Trends in Air Pollution Research
James, V. Livingston (Editor)
2005. ISBN: 1-59454-326-7

Agriculture and Soil Pollution: New Research
James, V. Livingston (Editor)
2005. ISBN: 1-59454-310-0

Water Pollution: New Research
A.R. Burk (Editor)
2008. ISBN: 1-59454-393-3

Air Pollution: New Research
James, V. Livingston (Editor)
2007. ISBN: 1-59454-569-3

Air Pollution Research Advances
Corin G. Bodine (Editor)
2007. ISBN: 1-60021-806-7

Marine Pollution: New Research
Tobias N. Hofer (Editor)
2008. ISBN: 978-1-60456-242-2

Complementary Approaches for Using Ecotoxicity Data in Soil Pollution Evaluation
M. D. Fernandez and J. V. Tarazona
2008. ISBN: 978-1-60692-105-0

Complementary Approaches for Using Ecotoxicity Data in Soil Pollution Evaluation
M. D. Fernandez and J. V. Tarazona
2008. ISBN: 978-1-60876-411-2 (Online Book)

Lake Pollution Research Progress
Franko R. Miranda and Luc M. Bernard (Editors)
2008. ISBN: 978-1-60692-106-7

Lake Pollution Research Progress
Franko R. Miranda and Luc M. Bernard (Editors)
2008. ISBN: 978-1-60741-905-1 (Online Book)

River Pollution Research Progress
Mattia N. Gallo and Marco H. Ferrari (Editors)
2009. ISBN: 978-1-60456-643-7

Heavy Metal Pollution
Samuel E. Brown and William C. Welton (Editors)
2008. ISBN: 978-1-60456-899-8

Cruise Ship Pollution
Oliver G. Krenshaw (Editor)
2009. ISBN: 978-1-60692-655-0

Water Purification
Nikolaj Gertsen and Linus Sønderby (Editors)
2009. ISBN: 978-1-60741-599-2

Environmental and Regional Air Pollution
Dean Gallo and Richard Mancini (Editors)
2009. ISBN: 978-1-60692-893-6

Environmental and Regional Air Pollution
Dean Gallo and Richard Mancini (Editors)
2009. ISBN: 978-1-60876-553-9 (Online Book)

Industrial Pollution including Oil Spills
Harry Newbury and William De Lorne (Editors)
2009. ISBN: 978-1-60456-917-9

Traffic Related Air Pollution and Internal Combustion Engines
Sergey Demidov and Jacques Bonnet (Editors)
2009. ISBN: 978-1-60741-145-1

Sludge: Types, Treatment Processes and Disposal
Richard E. Baily (Editor)
2009. ISBN: 978-1-60741-842-9

From Soil Contamination to Land Restoration
Claudio Bini
2010. ISBN: 978-1-60876-853-0

AIR, WATER AND SOIL POLLUTION SCIENCE
AND TECHNOLOGY SERIES

FROM SOIL CONTAMINATION TO LAND RESTORATION

CLAUDIO BINI
DEPT OF ENVIRONMENTAL SCIENCES
UNIVERSITY OF VENICE, ITALY

Nova Science Publishers, Inc.
New York

For permission to use material from this book please contact us:
Telephone 631-231-7269; Fax 631-231-8175
Web Site: http://www.novapublishers.com

Library of Congress Cataloging-in-Publication Data

Bini, Claudio.
Soil contamination to land restoration / Claudio Bini.
p. cm.
Includes index.
ISBN 978-1-60876-853-0 (softcover)
1. Phytoremediation. 2. Soil remediation. 3. Soil pollution. 4. Soil amendments. 5. Reclamation of land. I. Title.
TD878.48.B56 2009
628.5'5--dc22

2009048923

Published by Nova Science Publishers, Inc. ✝ New York

CONTENTS

PREFACE

Remediation of contaminated soils is one of the most important environmental issues. Chemical soil degradation affects 12% of all degraded soils in the world, totalling 2 billions hectares. Soil contamination is not only a social and sanitary issue, but has also an economic concern, since it implies major costs related to decreasing productivity and monetary evaluation of the contaminated sites. Costs related to remediation of contaminated soils (particularly with heavy metals), moreover, are very high.

Many of the organic substances contribute to contaminate ecosystems and are very poisonous to living organisms and to human health. Correspondingly, many metals, when present at high concentration in the environment, are critical or toxic to plants and animals, and may enter the food chain and therefore affect humans.

In areas affected by high contamination, direct and indirect health hazards require urgent restoration, regardless of the remediation technology selected for the site. In other cases, such as land with non-hazardous contaminant levels, remediation may eliminate or reduce the environmental hazard and contribute to the valorisation of green areas, public services, and arable land otherwise not utilizable.

Metal contamination persistence and little knowledge of mechanisms regulating the interaction soil-metal and the sorption of contaminants by living organisms make soil remediation particularly difficult and expensive. Any of the current technologies are actually effective and applicable at wide scale. The most utilized technical solutions are clearly inadequate for cleaning large areas of moderately contaminated land, where soft and (environmental) friendly technologies are needed to restore soil fertility, in such a way that they could be utilized for agriculture or public/residential green areas. Therefore, in recent years the interest of both public Authorities and private Companies towards innovative methodologies for decontamination and restoration of contaminated sites is increasing.

Phytoremediation is an emerging technology that holds great potential in cleaning up contaminants that: 1) are near the surface, 2) are relatively non-leachable, 3) pose little imminent risk to human health or the environment, and 4) cover large surface areas. Moreover, it is cost-effective in comparison to current technologies, and environmental friendly.

Most of the available data, until now, has come from microcosm experiments; full scale experiments could help in assessing the feasibility of phytoremediation , and its effective contribution to clean-up contaminated soils. However, phytoremediation is not yet ready for full scale application, despite favourable initial cost projections, which indicate expansion of clean-up market to be likely in next years.

Research should be addressed to find out new highly efficient accumulator plants, and related cultivation technologies, and this research must account for the spatial and temporal variability of complex systems that include mixtures of contaminants and organisms.

Chapter 1

INTRODUCTION

Soil and environmental contamination is a concern whose importance has been perceived only since recent years, and constitutes one of the great emergencies of XXI century, also because modern society is paying increasing attention to its effects on the human health, and is acquiring more and more consciousness of the disease risk connected to exposition to chemicals and toxic products like heavy metals, uranium, radionuclides, asbestos, benzene, dioxins, PCB, PAH. A demonstration is the increasing number of legal actions against public and private companies that are regarded as responsible for diseases or even death of workers (for instance, militaries who participated with the NATO army in the recent Bosnia-Serbia conflict are still dying by different cancer forms connected to exposition to impoverished uranium).

In the most part of industrialized countries, the problem of characterizing contaminated sites and their cleaning up is increasingly relevant in soil and environment safeguarding, also due to the augmented population sensitivity. Areas previously occupied by highly contaminant industries, like power plants, fuel refineries, smelters, tannery plants, present high contamination levels by both organic and inorganic substances. Many of the organic substances which contribute to ecosystem pollution are highly noxious to human health and to living organisms. Similarly, many metals that are present in the environment at determined concentration levels, may enter the food chain, and be critical or toxic to living organisms and humans. Risk assessment for human health, therefore, is assuming increasing importance in the solution of problems connected to soil clean up and to land restoration. It is utilized, in fact, to identify and classify sites on the basis of the intervention priority, to establish decontamination objectives and standards, to select the proper technology for each specific situation.

Direct and indirect health risks make urgent to clean up areas highly polluted, and acceptable the costs and the investments to sustain, irrespective of the strategy selected for restoration. In other cases of less gravity, like soils having not hazardous metal contamination levels, or when costs would be excessive with respect to the estimated benefits, intervention may eliminate or reduce environmental hazard, allowing restoration of degraded land and their valorisation as green areas, public services, productive utilization, thus favouring the establishment of an actual business in the sector of environmental restoration.

Chapter 2

BACKGROUND AND LEGISLATIVE SOIL REFERENCE VALUES

Several definitions of contaminated sites are given elsewhere. A site is contaminated when it presents chemical, physical or biological alterations of soil, or subsoil, or surface water, or groundwater, in such a way that a danger to public health, or to the natural, or constructed environment may arise. It may be of natural, or anthropic origin.

Natural contamination is related to geochemical anomalies connected to geological factors (e.g. rock materials and minerals enriched in metals, like Ni, Cr, Cu in serpentine, As in fossil flower) or to mining areas and ore deposits (e.g. toxic metal sulphides like Ag, Cu, Pb, Zn, Hg). Mine dumping constitutes a further problem, since, besides the metal "hot spots", diffuse land and water contamination may occur. Spreading of mined material over large areas (Figure 1) originates mine dumps which are enriched in phytotoxic metals, and therefore highly infertile; moreover, the restoration of such areas may require elevated costs.

Contamination of anthropic origin, instead, is related to the presence and accumulation of contaminants originated by human activities, including urban waste disposal (Figure2), and therefore is more important and worrying, since it is diffused worldwide. Industrial activities are the main causes of pollution, although at localized hot spots, whereas agriculture is responsible for diffuse contamination. One of the most recent issues is the pollution caused by metallic fragments introduced into soil because of war activities (Souvent and Pirk , 2001; Van Meirvenne et al., 2008), including damage to living organisms and humans by impoverished uranium.

Figure 1. Mine dumps discharged in the close vicinity of an abandoned mixed sulphide mine highly contribute to soil infertility and random vegetation. (photo Bini, 2006).

Figure 2. Anthropogenic soil developed on urban waste disposal. (courtesy C. Dazzi).

Another source of important soil contamination is atmospheric deposition caused by industrial emissions, motor vehicles, acid rains, etc. (Bini, 2008a).

A list of the most significant activities, in terms of contamination,includes:

- industrial activities (petrol, chemicals, metallurgy, varnish, tannery, electronics…);
- emissions and discharge (power plants, motor vehicles, fossil fuel…);
- composting; urban solid residues and waste; landfills;
- agriculture (fertilizers, pesticides, sewage sludge…).

Contaminants may be distinguished, according to their composition and nature, in two categories with different diffusion, health hazard and remediation technology: organics and inorganics.

The main organic contaminants are:

- mineral oil (fossil fuel, gasoline, diesel, lubricants…);
- aromatic compounds (PAH, PCB…);
- combustion products (dioxins…);
- agrochemicals.

Inorganic contaminants are:

- Heavy metals (Cd, Cr, Ni, Cu, Zn, Pb…);
- Light metals (Al, Be, Tl, F, Br…);
- volatiles (As, Hg, Se);
- radionuclides (Cs, U, Ra…);
- anions (nitrates, nitrites, phosphates…).

Concerning soil, in particular, a soil is contaminated[1] when its concentration of contaminants exceeds the background level. Background level corresponds to the total content of metals in soils not affected by human activities. These values are available in a number of publications (Alloway, 1995; Tobias et al., 1997; Adriano, 2001; Baize and Sterckeman, 2004; Reiman and Garret, 2005).

---

1 Contamination occurs when the soil composition deviates from the normal composition. In their natural state contaminants may not be classified as pollutants unless they have some detrimental effects to the organisms. Pollution occurs when a substance is present in greater than natural concentration as a result of human activities and having a net detrimental effect upon the environment and its components (Adriano et al. 1995).

Background values may vary as a function of the locality from which a given soil is sampled. For example, metal concentrations in serpentine-derived soils can be highly toxic to animals and plants as a result of the naturally elevated metal contents of the parent rock from which the soil is derived. Similarly, metal concentrations in soils are known to be affected by the clay content of soils and increase almost linearly as a function of it (Jenny, 1941). Organic matter, in turn, may determine metal behaviour in the soil-plant system, and therefore the possible translocation to plants (bioavailability).

Because of the different forms and the spatial variability of metals in soils, background values do not serve as good reference values for legislative purposes. Therefore, it is not possible to arrive at a single background value for any of the metals. In an effort to expedite remediation of hazardous waste sites in the absence of a national soil cleanup standard, many National Agencies have developed their own clean-up standards. In general, cleanup levels promulgated for industrial sites tend to be up to one order of magnitude less stringent than those for residential sites. On the other hand, soil clean-up levels established to protect groundwater quality tend to be more stringent than those established based on direct human exposure to contaminated soils. In addition, carcinogens tend to be assigned more stringent levels than non-carcinogens. The USEPA (1993) has proposed a classification scheme for carcinogenity based on human evidence. Substances in Group A are *known* human carcinogens (e.g. radon, dioxins, vinyl choride, benzene), Group B refers to *probable* human carcinogens (e.g. As, Cd, Cr, Hg), Group C refers to a *possible* carcinogen (e.g. N, U), Group D refers to unclassified substances because of inadequate data (e.g. Thallium), and Group E refers to substances with evidence of noncarcinogenicy. Although no U.S.A. federal levels have been developed for regulatory purposes of hazardous constituents in soil, health risk-based soil screening levels were drafted by the US-EPA in autumn of 1993 (Bryda and Sellman, 1994). These levels are used to assist in the assessment of the maximum contaminant level (MCL) of soils at sites that pose potential concern, as well as screen out those soils that do not request additional actions. Cleanup levels developed for metals in the U.S.A. are based on average background concentrations found in soils or in standard risk assessment methods (Bryda and Sellman, 1994).

Some other countries, notably Canada, Great Britain, Belgium and the Netherlands have progressed further in setting up soil standards for soil remediation. In the Netherlands, the discovery of a contaminated residential area created a public complaint that led to a legislative mandate for soil restoration in this country. Metals, inorganics, and a wide range of organic compounds were involved (Table 1).

Table 1. Provisional estimation of the health risk connected to different contaminants (source: Van Hall Intituut Groningen, The Neederland, 1998)

Contaminant	Source	Exposure routes	Health risk
As, Cd, Cr, Hg, Pb, Ni, Cu, Zn	Industrial activities (varnish,battery,steel…), combustion	Inhalation, ingestion, dermal contact, food chain	Carcinogenic, teratogenic,mutagenic, phytotoxic
Nitrates, nitrogen oxides	Chemical industry	Ingestion, inhalation	Toxic; carcinogenicy unclear
Dioxins and related compounds	Combustion processes	Ingestion, food chain	Very toxic, carcinogenic
PAH	Fuel, storage tanks	Inhalation, ingestion, dermal contact	Toxic to nervous system; carcinogenic
Chlorinated hydrocarbons, Organochlorinated pesticides	Chemical industry; Petrol industry, agrochemicals	Inhalation, ingestion, dermal contact, food chain	Toxic; carcinogenic

For each contaminant, three different values were initially adopted:

A. Mean reference value;
B. Threshold value for pollution, above which no biological or ecological damage is yet observed; soils with this level of pollution, however, should be further monitored.
C. Threshold value above which restoration is recommended.

These criteria were recently revised. Table 2 presents the values for metals adopted by the Dutch Legislature. The intervention values for soil remediation will be used to assess whether contaminated land poses serious threat to public health. These values indicate the concentration levels of the metals in soil above which the functionality of the soil for human, plant, and/or animal life is seriously compromised or impaired. Concentrations in excess of the intervention values correspond to serious contamination. The intervention values replace the old C values in the soil protection guidelines.

Table 2. Dutch target values (also referred to as A-value or reference value) and intervention values (also referred to as C-value) for selected metals for soil (mg/kg dry matter).(Source: Dutch Ministry of Housing, Spatial Planning and Environment. The Hague, The Netherlands)

Metal	target value	intervention value
Arsenic	29	55
Barium	200	625
Cadmium	0.8	12
Chromium	100	380
Cobalt	20	240
Copper	36	190
Mercury	0.3	10
Lead	85	530
Molybdenum	10	200
Nickel	35	210
Zinc	140	720

The recorded values,

- are based not only on considerations of the natural concentrations of the contaminants which indicate the degree of contamination and its possible effects, but also of the local circumstances, which are important with regard to the extent and scope for spreading or contact;
- are related to spatial parameters. The soil is regarded as being seriously contaminated, if the metal mean concentration in at least 25 cubic meters of soil volume exceeds the intervention values;
- are dependent on soil type, since they are related to the content of organic matter and clay in the soil.

The target values (Table 2) are important for remedial as well as for preventive policy. They indicate the soil quality levels ultimately aimed for a given utilization. These values are derived from the analysis of field data from relatively pollution-free rural areas regarded as non contaminated, and take into account both human toxicological and ecotoxicological considerations.

In other European countries, the public has asked for a similar legislation for soil restoration. Regulatory guidelines on tolerable metal concentrations for agriculture and horticulture were published in Germany (Kloke et al., 1980) and in Switzerland (Vollmer et al., 1995). In the U.K., different land use categories were proposed as a criterion for determining the threshold value for development of contaminated sites, after restoration (guidance 59/83, Department of the Environment, London, 1987). In Germany, Eikmann and Kloke (1993) introduced threshold values for playgrounds, parks, parking areas, and industrial sites. In agricultural and horticultural soils, lower threshold values were proposed when growing leafy vegetables than for fruit production or for the cultivation of grain or ornamental plants (Eikmann and Kloke, 1995). A similar approach has been proposed in Poland where agricultural and horticultural uses vary according to the severity of soil contamination (Kabata-Pendias, 1997).

In the recent environmental legislation of Belgium, the threshold values for restoration that are somewhat corresponding to the intervention values vary with the intended land use for the remediated site. These threshold values were defined using the "Human Exposure to Soil Pollution Model" by Stringer (1990), which estimates the transfer of contaminants from soil to man by different pathways (i.e., by inhalation, ingestion, drinking water, animal or plant food, etc.). It was recently improved and several other models are proposed to assess the human risk of soil pollution.

In Italy, following the EU Directive on Soil Protection, the current regulatory process has been recently revised with the Legislation Act n°152/2006, which indicates the criteria for identifying contaminated sites, suggests possible safety and restoration interventions, and introduces the concept of *risk threshold concentration* and *contamination threshold concentration*. Practically, a site is contaminated when the concentration of just one of the contaminants overpasses the risk threshold values reported in the contaminants regulatory list. A provisional list of admissible contaminant concentrations for green areas, residential and industrial sites is reported in Table 3.

Table 3. Maximum concentration values recordable in soil and subsoil of contaminated sites, with reference to specific land utilization

Chemicals Inorganic compounds		Green and residential areas mg/kg d.m.	Commercial and industrial areas mg/kg d.m.
	Antimony	10	30
	Arsenic	20	50
	Berillium	2	10
	Cadmium	2	15
	Cobalt	20	250
	Chromium (total)	150	800
	Chromium VI	2	15
	Mercury	1	5
	Nickel	120	500
	Lead	100	1000
	Copper	120	600
	Selenium	3	15
	Tin	1	350
	Thallium	1	10
	Vanadium	90	250
	Zinc	150	1500
	Cianides	1	100
	Fluorides	100	2000
Organic compounds			
	Benzene	0.1	2
	Ethylbenzene	0.5	50
	Styrene	0.5	50
	Toluene	0.5	50
	Xylene	0.5	50
	Benzo(a)antracene	0.5	10
	Benzo(a)pyrene	0.1	10
	Benzo(b)fluorantene	0.5	10
	Benzo(k) fluorantene	0.5	10
	Crisene	0.5	50
	Dibenzo(a)pyrene	0.1	10
	Dibenzo(a,h)anthracene	0.1	10
	Indenopyrene	0.1	50
	Pyrene	5	10

Chapter 3

SOIL REMEDIATION AND RISK ASSESSMENT

Remediation of contaminated soils is one of the most important environmental issues. Chemical soil degradation affects 12% of all degraded soils in the world, totalling 2 billions hectares (E.U. Commission, 2006). Soil contamination is not only a social and sanitary issue, but has also an economic concern, since costs related to remediation of contaminated soils (particularly with heavy metals), are very high. Therefore, only few developed countries (USA, G.B., The Netherlands, Germany, Australia) have started remediation actions, whereas many developing countries do not yet have started remediation projects, although they are affected by high environmental hazards (e.g. As in soils and groundwater in Bangladesh; U in soils of Bosnia, as a consequence of the recent civil war).

In the USA, the remediation of the sites listed in the National priority List in 1986 (40% of the whole) would account for 7 billions $ (Salt et al., 1995), and more than 35 billions $ are accounted for the remediation of the over 1000 sites which have been identified as hazardous.

In Switzerland, 10,000 ha of arable land have Zn concentration above the target value, and 300,000 ha present high levels of Cd, Pb and Cu (Vollmer et al.,1995).

A research carried out in five European Union countries (Table 4) allowed identification of more than 22,000 contaminated industrial sites in critical conditions (totally 0.2% of the land), for which an immediate intervention is required to safeguard public health, or have severe limitations in their utilization, and more than 50,000 sites need further investigation in order to assess their actual hazard.

Table 4. Number of contaminated sites in selected European Union countries (Adriano et al., 1995)

Country	Contaminated Sites (total)	Sites in crytical conditions
Germany	32,000	10,000
Belgium	8,300	2,000
Italy	5,600	2,600
Netherland	5,000	4,000
Denmark	3,600	3,600

The Italian Environmental Agency estimates that at present (2005) contaminated areas which need remediation overcome 10,000 sites. Of these, areas previously settled by highly contaminating factories (e.g. chemicals at Porto Marghera, Venice; metallurgy at Bagnoli, Naples; tannery factories at Arzignano, Vicenza and S. Croce, Pisa) present very high contamination levels by organic as well as inorganic substances. A priority list of intervention indicates that the petrol-chemical area close to the lagoon of Venice (Figure 3) is the major contamination concern in Italy. Deep investigation on trace elements in soils, sediments and water of the lagoon watershed (Zonta et al., 2007; Bini, 2008a) revealed heavy contamination, and suggested a master plan aimed at decontamination of the lagoon and the conterminous land.

Soil contamination, as previously stated, is also an economic concern, since it implies major costs in terms of soil fertility loss, agricultural products worsening, and ultimately monetary evaluation decrease. Estimates related to the last decades indicate that contamination, besides erosion, is the main cause of soil loss, accounting for more than 3ha soil/min each year lost by contamination (Bini, 2008b). The quantified economic losses would amount to more than 3 billions $/year (Pierzynski, 2003).

Many of the organic substances (PCB, PAH, etc.) contribute to contaminate ecosystems and are very poisonous to living organisms and to human health. Correspondingly, many metals, when present at high concentration in the environment, are critical or toxic to plants and animals (Salomons, 1995), and may enter the food chain and therefore affect humans. Concerning particularly heavy metals, at worldwide level it is estimated that approximately 1 billion persons is affected by Pb contamination disease, approx 500,000 by Cd, more than 100,000 by As, with an annual addition to soil of about $98x10^3$ kg As year^{-1} (Ungaro et al., 2008), without considering the Asian countries – Pakistan,

Bangladesh, Korea – where water pollution by As is dramatic, and population health risk is very high.

Figure 3. An overview of the contaminated site at Porto Marghera, Venice (Italy). Petrol-chemical plants are considered the main responsible for the Venice lagoon pollution. (Bini, 2004).

The risk assessment for human health, therefore, is assuming more and more importance in the solution of problems connected with soil remediation. Indeed, the risk assessment criteria are applied to identify and classify the various sites on the basis of intervention priority, to establish objectives and standard of decontamination, to select the technology more appropriate and site-specific.

Risk assessment is defined as the process of estimating the probability of occurrence of an event and the probable magnitude of adverse health effects over a specified period of time (Lee et al., 2008; Lim et al., 2008). Human health risk assessment consists of four stages: 1) hazard identification; 2) dose-response (toxicity) assessment; 3) exposure assessment; 4) risk characterization and quantification.

The purpose of hazard identification is to identify chemicals which may have a harmful effect in human body. A hazard is a source of risk, but not a risk itself.

The purpose of toxicity assessment is to estimate the potential for selected chemicals to cause harmful effect in exposed people, and to provide an estimate of the relationship between the extent of exposure and the increased probability of harmful effects. The principal toxicity index for non cancerogenic effects is the reference dose (RfD), i.e. the estimated amount of the daily exposure level for the population that is likely to be without an appreciable risk of deleterious effects during a lifetime. The slope factor (SF) for cancerogenic effects is the probability of an individual developing cancer as a result of unit daily intake exposure over a lifetime.

The exposure assessment is the evaluation of exposure routes and pathways of a receptor. The average daily dose (ADD) is the quantity of chemicals ingested, inhaled or absorbed per kilogram of body weight per day (mg/kg/day).

The risk characterization is a quantitative estimation of cancer risk and hazard index (HI) for multiple substances:

Cancer risk= ADDxSF
Hazard Index= ADD/RfD.

The acceptable cancer risk for regulatory purposes is in the range of 10^{-6} – 10^{-4} i.e. one case in a population ranging from 1 million to 10,000 people is acceptable. If the calculated HI is less than 1.0, the non-carcinogenic adverse effect due to a given exposure pathway is assumed to be negligible.

In areas affected by high contamination, direct and indirect health hazards require urgent restoration and acceptable costs, regardless of the remediation technology selected for the site. In other cases, such as land with non-hazardous contaminant levels, or excessive costs compared to the expected benefits, remediation may eliminate or reduce the environmental hazard and contribute to the valorisation of green areas, public services, and arable land otherwise not utilizable.

Decision makers should evaluate the selection of the remediation technologies also in relation to the effects that it may have on the soil quality. Many processes, indeed, determine significant changes in soil characteristics (e.g. pH variation, red-ox conditions, fertility, structure loosening, sterilization and decline of biological activity). Action for restoration of degraded areas, therefore, should take care of both costs for remediation and management of the site to secure, of the hazards derived from the site itself, and of the benefits derived from site restoration.

Metal contamination persistence and little knowledge of mechanisms regulating the interaction soil-metal and the sorption of contaminants by living organisms make soil remediation particularly difficult and expensive. Any of the current technologies are actually effective and applicable at wide scale. The most utilized technical solutions are clearly inadequate for cleaning large areas of moderately contaminated land, where soft and (environmental) friendly technologies are needed to restore soil fertility, in such a way that they could be utilized for agriculture or public/residential green areas. Therefore, in recent years the interest of both public Authorities and private Companies (e.g. Dupont, Monsanto) towards innovative methodologies for decontamination and restoration of contaminated sites is ever increasing.

The risks associated with polluted soils vary from site to site according to scientific database, public perception, political perception, national priority, etc. While severely contaminated soils may require some form of remediation, there may be instances where remediation is not desirable (Adriano et al., 1995). These include:

1. the cost of clean-up far exceeds the expected benefits of clean-up in terms of human health and ecological sustainability;
2. the contaminated soil is not being used and has a low potential to be used in the future;
3. there are inexpensive substitutes for the contaminated soil in question;
4. the site will not be used after remediation because users will take some averting action, and
5. the contamination does not degrade soil and/or water quality to an unsafe or unhealthy level (NRC, 1993).

In the U.S.A., a recommended systematic procedure for remedial action is based on the following items (Adriano et al., 1995):

1. reporting and identification;
2. selection of response action;
3. preliminary assessment/site investigation;
4. remedial investigation/feasibility study;
5. remedial design/remedial action;
6. operation and maintenance/post closure monitoring.

In arriving at a remedial decision, there are three categories of criteria that must be considered according to the National Contingency Plan (Grasso, 1993):

- Threshold criteria:

Overall protection of human health and the environment;
Compliance with applicable or relevant and appropriate requirements;

- Primary balancing criteria:

Long-term effectiveness and permanence;
Reduction of toxicity, mobility, or volume through treatment;
Short-term effectiveness;

Implement ability;
Cost;

- Modifying criteria:

State acceptance;
Population consensus.

The final choice of remedial technology largely depends on the nature and degree of contamination, the intended function or utilization of the remediated site and the availability of innovative and cost-effective techniques. The choice is further complicated by environmental, legal, geographical, and social factors. More often the choice is site-specific. For example, home gardens and agricultural fields in large rural areas that are contaminated may require a remedial approach different from that for smaller but heavily contaminated areas. Similarly, large areas around old mining and smelter sites need an approach which differs from that of a heavily polluted spot.

3.1. Methods of Soil Remediation

The methods and techniques for remediating contaminated soils may be subdivided into two strategies:

- confination;
- treatment.

Confination technologies include (civil) engineering techniques that have the objective of removing or isolating the source of contamination, or of modifying migration ways or percourses. Such techniques comprehend:

- excavation and landfilling both inside and outside the site;
- barriers created in the contaminated soil;
- soil incapsulation; soil solidification;
- hydraulic intervention (pumping, washing).

Important factors driving the selection of such techniques are:

- large space available within the contaminated land;
- available geological and hydrogeological background studies;

- availability of natural/seminatural materials (geomembranes, geotextiles) to dress the excavated materials and to cover the contaminated material;
- the possible impact derived from excavation and /or disturbance;
- sterilization of the whole area devoted to infrastructures and building constructions.

None of these techniques are entirely satisfactory (Exner, 1995). Landfilling is a temporary solution that delays remediation. Furthermore, it has been discontinued in most countries. Incapsulation/solidification does not remove the contaminant from the soil, thereby greatly limiting the value of the soil. Soil washing and flushing have been used extensively in Europe but only had limited use in the U.S.A. The process involves excavation of the contaminated soil, mechanical screening to remove various oversized materials, separation processes to generate coarse- and fine-grained fractions, treatment of those fractions, and management of the generated residuals. Soil washing performance is closely tied to three key physical soil characteristics: particle size distribution, contaminant distribution among the different size particles, and how strongly the soil binds the contaminant. In general, soil washing is most appropriate for soils that contain at least 50% sand and gravel, such as coastal sandy soils and soils with glacial deposits (Westinghouse Hanford Co., 1994).

Treatment technologies are based on processes addressed to removal, stabilization or destruction of contaminants.

- Removal may be attained by contaminant mobilization and/or accumulation processes (leaching, sorption), contaminant concentration and recovery processes (physical separation) or a combination of processes (accumulator plants).
- In-situ stabilization consists of the contaminant being made less mobile and therefore less toxic by a combination of physical, chemical and biological processes.
- Contaminant destruction is achieved by physical, chemical or biological degradation (e.g. thermic or microbiological treatments).

Treatment processes may be operated according to their application, namely:

- ex-situ, when operated in the area of the contaminated site;
- in-situ, when they are operated without removing the contaminated soil;
- on-site, when treatment is operated in the area of the contaminated site, by moving and removing the contaminated material;

- off-site, when contaminated material is moved from the site, and transported to the treatment plants or to landfill.

Chemical treatments involve contaminants destruction or removal by:

a) oxidation (change to higher chemical valence): many organic compounds, for instance, are oxidized to CO_2;
b) reduction (change to lower chemical valence): CrVI may be reduced to CrIII, which is less mobile and less toxic than CrVI;
c) immobilization: contaminant mobility is reduced through precipitation as an insoluble complex, or adsorption on solid matrices, etc.;
d) extraction: contaminant is extracted from soil material by application of different extractants (organic compounds, acids, tensioactives, etc); percolating liquid may be collected and treated for more degradation, or sent to landfill;
e) substitution: some chemical groups of the contaminant may be substituted with other groups, that make the contaminant less toxic (e.g. dealogenation of chloride solvents).

Physical treatments are aimed at separating contaminants from the soil matrix, taking into account the differences between contaminant and soil characteristics (e.g. volatility, magnetic properties, density, etc). They include several processes like electrokinetic, electrolysis, electroosmosis, electrophoresis, stripping extraction, etc.

Thermic treatments utilize elevated temperatures to prime physical and chemical processes like volatilization, ash flying, pyrolysis, etc, thus allowing contaminant removal or destruction, or immobilization in the soil matrix. Two main thermic treatments are available:

- desorption (working temperature is in the range 100°C – 800°C), and
- incineration (working temperature is in the range 800°C – 2500°C; contaminant is destroyed).

Once a contaminant is volatilized from the soil, it may be successively removed from the gas-phase through condensation or combustion. High-temperature incineration is the most common method of dealing with metal-contaminated soils because it has been proven to be the most reliable destruction method for the broadest range of wastes (Lee et al., 1990). However, it is costly and is often difficult to obtain a legal permit for it.

Biological treatments are known also as "bioremediation", i.e. "utilization of living organisms to reduce or eliminate environmental contaminants" (Adriano et al., 1999). Biological treatments include one or more of the following processes:

- degradation: contaminant biochemical degradation by soil microorganisms (bacteria, fungi, actinomycetes);
- transformation: contaminant biochemical conversion to make it less toxic and /or less mobile;
- accumulation: organic and inorganic contaminants may accumulate in tissues of living organisms (particularly plants);
- mobilization: a contaminant-bearing solution may be separated form the contaminated soil.

Microorganisms are potentially able to detoxify several contaminated sites, and to bring them back to the original state. Higher plants are utilized to stabilize or remove contaminants (especially heavy metals) from soil and waters. This technology, known as phytoremediation, is potentially little destructive, environmental friendly and cost-effective, and is applicable to large contaminated land.

Phytoremediation is a technique which utilizes plants to remove, eliminate, or decrease environmental contaminants (heavy metals, organics, radionuclides, explosives), attenuating the related risk (Barbafieri, 2001; Li et al., 2002). It is based on several natural processes which involve plants:

- direct sorption of metals or moderately hydrophobic organic compounds;
- accumulation or transformation of chemicals by lignification, metabolization, volatilization;
- catalysis and degradation of organic compounds by enzymes released by plants;
- exudates release in the rhizosphere, with pH modification, carbon and microbial activity increase, and contaminant degradation.

Below are some techniques utilized in phytoremediation for removing inorganic contaminants from soils based on the above processes, that the clean-up industry is already utilizing or seriously considering to adopt.

Phytostabilization

Phytostabilization is a process in which plants tolerant to contaminant metals are used to reduce the mobility of contaminant metals, thereby reducing the risk of further environmental degradation by leaching into the groundwater or by airborne spread (Smith and Bradshaw, 1979; Losi et al., 1994; Vangronsveld et al., 1995a). In *in situ* metal stabilization, soil amendments can be combined with the use of metal-tolerant plants to enhance plant growth and stabilizing effect of the treatment (Vangronsveld et al., 1995a; 1995b; 1996). Metal-tolerant plants immobilize contaminants at the interface root-soil by absorption, precipitation or complexation, thus reducing mobility and migration to groundwater, or to the food chain, as is the case of chromium (Bini et al., 2000; 2008c).

Stabilization may occur:

- in the root zone: proteins are released (by roots) in the rhizosphere, and determine precipitation of contaminants;
- on cell walls: proteins associated with root cell walls may stabilize contaminant outside the root cell, thus impeding contaminant to be transported inside the plant (barrier effect);
- in root cells: proteins present on root membranes may enhance contaminant transport inside the cell, where it is sequestered in vacuoles.

Arboreal and shrubby plants plants seem to be more prone than herbaceous ones to apply phytostabilization, owing to different sorption ways and metal mobility in the plant organs. Some ornamentals, like bay (*Laurus nobilis*), pitosphor (*Pitosphorum tobira*), oleander (*Nerium oleander*) proved effective in metal stabilization (Carratù et al., 2001). Among the herbaceous plants, monocotyledons, both native (e.g. *Lolium perenne*) and cultivated (e.g. *Zea mays*, *Avena sativa*) are more suitable than dicotyledons (Argese et al., 2001).

Phytostabilization is particularly suitable at sites where it is important to keep metals in non-mobile form, in order to impede dispersion, like it happens for chromium (Bini et al., 2000a; 2008c). Moreover, when metal concentration is very high, phytoextraction would require long time to achieve the objectives of land restoration.

Rhizofiltration is a process in which plant roots absorb, precipitate and concentrate heavy metals from polluted wastewater streams (Dushenkov et al., 1995). It proved particularly effective in taking up As from wetland areas (Sette et al., 2001).

Phytostimulation

This technique concerns stimulation of microbial or fungal degradation by means of exudates and enzymes in the rhyzosphere (Schnoor et al., 1995). Root exudates (organic acids, alcohols, sugars) have a stimulating effect on microbial activity, thus increasing the biodegradation capacity of bacteria and fungi, as observed by Mehmannavaz et al. (2001) with *Sinorhizobium meliloti* on PCB in the rhizosphere.

Phytostimulation is a symbiontic relation between plants and soil microorganisms. The former provide nutrients to microorganisms, and the latter determine soil decontamination and favour root development.

Phytoextraction

Several plants show a marked ability to accumulate contaminants (heavy metals, radiactives) in their aerial parts, and for this reason are known as "hyperaccumulator plants" (Wenzel et al., 1993; Baker et al., 2000). At the end of the phoenological cycle (or when the metal sorption is concluded), the plant may be harvested, and contaminant recovered by distillation. Hyperaccumulator plants are able to accumulate metals up to 500 times higher than metal concentration in non-accumulator plants (Lasat, 2002), with metal concentration in leaves even higher than 5% dry weight (Mc Grath, 1998). Hyperaccumulator plants may accumulate more than 10 mg/kg Hg, 100 mg/kg Cd, 1,000 mg/kg Cu, Cr, Pb, and 10,000 mg/kg Zn. More than 400 species of hyperaccumulator plants are known at present (Baker et al., 2000), and are subdivided in 45 families, of which the most represented is Brassicaceae (Marchiol et al., 2004). Not all the metals are accumulated in plants in the same way and to the same extent: some plants absorb only one metal, some others more metals; for some metals, like thallium, there are not yet known accumulator plants; for some other metals, like arsenic, some species, like fern (*Pterix vittata*) have been discovered recently (Ma et al., 2001; Tu et al., 2004).

As a general rule, metals like Cu, Cd, Zn, Ni, Pb, Se are easily accumulated, whilst As, Co, Cr, Mn, Fe, U are difficult to accumulate.

Phytoextraction efficiency depends upon several factors:

- the nature and concentration of contaminant (red-ox status, binding, bioavailability, etc.);
- soil/sediment chemical-physical characteristics (pH, texture, CEC, etc);

- morphological and physiological plant characteristics (root pattern, absorption capacity, metal synergism or antagonism, etc).

A coefficient currently utilized to evaluate phytoextraction efficiency is the Biological Absorption Coefficient (BAC = (metal)plant/(metal)soil; Ferguson, 1992). The total amount of removed contaminants results from the harvested plant biomass multiplied by the metal concentration in plant.

As already stated, once ceased the absorption and translocation of metal from roots to the aerial parts (phenological cycle, maximum metal concentration allowed, more metal unavailability, etc), plants may be harvested and transported to landfill; the contaminated site may be subjected to repeated cultivation cycles, in order to decrease metal concentration to acceptable levels, or at the best to eliminate it at all. Quite recently, has been explored the opportunity of recovering metals sorbed by plants by distillation (Cunningham and Berti, 1993; 1997). This technique proved interesting in an economic perspective: in the U.S.A., Canada, Australia and possibly in some more countries, some private Companies have bee established to economically utilize these low cost “ore outcrops” (Chaney et al., 1997). One of the most experimented plants is the known Zn-Cd hyperaccumulator *Thlaspi caerulescens*. However, the reduced plant biomass (which is common to many hyperaccumulator plants) does not allow recovery of relevant metal amounts. To overcome this important limiting factor, that would diminish the economic relevance of this technique, research in genetic engineering is in progress (Raskin, 1996; Mc Nair, 1997; Prasad, 2003), in order to produce plants with higher biomass, and therefore with major ability to remove metal from soil.

Assisted Phytoremediation

One of the major limits in soil decontamination with plants, as already stated, is the little biomass of most plants, especially those which grow in fresh to temperate climate. A second factor limiting the phytoextraction effectiveness is the soil metal bioavailability, i.e. the ability of plants to take up a metal from the soil. Bioavailability depends upon chemical and physical conditions, bacteria, fungi and plants that may influence such conditions (Ernst, 1996), upon organic complexes be formed, inorganic compounds precipitation, etc.. The assisted phytoextraction increases metal bioavailability by applying to soil syntetic chelatants like EDTA, HEDT, DTPA. The EDTA results the most effective in increasing Ni and Pb extraction by several plants (Kramer, 1996; Blaylock et al.,

1997). This technique, however, presents high environmental impact, since chelatants may be leached to subsoil and groundwater. Moreover, chelatant treatment may influence plant growth, if the phytotoxicity threshold is overpassed (McGrath, 1998). Recent alternatives to chelatants application are increasing microflora in the rhizosphere with microrganisms (Whiting, 2001) and amendants application (e.g. zeolites: Zaccheo et al., 2001), that make the soil more suitable, thus enhancing plant growth.

3. 2. Current Status and Perspectives in Phytoremediation

Phytoremediation is an emerging technique to clean up contaminated sites. It allows both organic and inorganic contaminants to be removed from soil and water, and physical stabilization of contaminated soils. It presents numerous advantages in comparison to other treatment techniques:

- improvement of chemical, physical and biological soil properties;
- erosion rate reduction;
- waste production reduction;
- potential metal recovery;
- land aesthetic improvement;
- high population consensus;
- reduced costs in comparison to other remediation technologies (40% reduction in respect to other in-situ technologies, and up to 90 % reduction in respect to ex-situ technologies).

Some disadvantages, however, affect this technology:

- it is strongly depending upon environmental and climatic conditions;
- strongly depending upon contaminant concentration and bioavailability;
- it depends on the contamination area extent and depth (it should be less than 5 m under the ground level);
- it needs longer time for land restoration, with respect to other technologies.

Although current literature on phytoremediation is rather abundant, the physiological mechanisms that control and regulate the above processes are not yet elucidated, also because many of these are site-specific. The specificity depends first of all upon the kind of contaminant, its composition and possible transformation (evolution), and, consequently, upon the vegetal species to be utilized for remediating a contaminated site. Further investigation, therefore, is needed in order to elucidate both the mechanisms involved and selection of plants, possibly applying genetic engineering to increase plant biomass.

Phytoremediation is particularly suitable at sites where contamination is rather low and diffused over large areas, its depth is limited to the rhizosphere or the root zone, and when there are no temporal limits to the intervention. It has been calculated, indeed, that with present accumulator plants, at least 3-5 years are needed to have appreciable results in clean up a moderately contaminated soil (McGrath, 1998). McGrath (1995) calculated that nine harvesting of *Thlaspi caerulescens* are necessary to decrease Zn concentration in soil from 444 to 300 mg/kg; conversely, a non-accumulator plant, like radish, needs 2046 harvesting cycles to attain similar results. The same *Thlaspi caerulescens* and the hyperaccumulator *Cardaminopsis halleri* may remove, within only one harvest, Cd accumulated in decades of years of phosphate fertilizer application, with a removal rate of 150 g/ha/y, and 34 g/ha/y, respectively (McGrath and Dunham, 1997).

One of the peculiar properties of phytoremediation is the economic aspect, since it presents costs much lower relative to the other technologies. Black (1995) estimated costs of only 80 $\$/m^3$ of soil cleaned with phytoextraction, and 250 $\$/m^3$ with soil washing. According to Cunningham and Berti (1997), costs for phytodepuration would be 100,000 $/ha, and excavation and landfilling would amount 500,000 $/ha, while costs for traditional technologies of in-situ treatment would range between 500,000 and 1,000,000 $/ha. However, actual costs for phytoremediation are still highly variable, and will be better estimated when this innovative technology will be really effective. Presently, indeed, the paucity of full scale application makes difficult to obtain reliable indication on remediation time and costs. A comparison of costs for different remediation technologies is reported in Table 5. It is noteworthy to point out that many ex-situ treatments require the combined use of different items reported in the table, and therefore the whole cost would be higher than the single item.

Future perspectives are relied to the remediation protocol set up by U.S.A., which ultimately allows for metal recovery and commercialisation, at decontamination cost decreased by at least 10 times (0.25 M$ vs 3M$) with respect to ex-situ decontamination, and at low environmental impact.

Table 5. Comparative costs of different remediation technologies per soil unit (adapted from Adriano, 2001)

In-situ treatment	Costs (U.S. $/m^3$)
Soil flushing	50-80
Bioremediation	50-100
Phytoremediation	10-35
Ex-situ treatment	
Exavation and transport to landfill	30-50
Disposal in sanitary landfill	100-500
Incineration or pyrolysis	200-1500
Soil washing	150-200
Bioremediation	150-500
Solidification	100-150
Vetrification	Up to 250

Numerous Companies in different countries of Europe, Australia, Canada and U.S.A.(e.g. GLASS, DUPONT) proved interested to this business.

An economically interesting strategy of phytoestraction has been proposed recently by Vangronsveld et al. (2007) for soil remediation. Selected food crops as rapeseed, maize and wheath are grown on contaminated soil, harvested and treated with different techniques, in order to obtain metal recovery and biogas production, according to the following scheme, adapted from Vangronsveld et al. (2007).

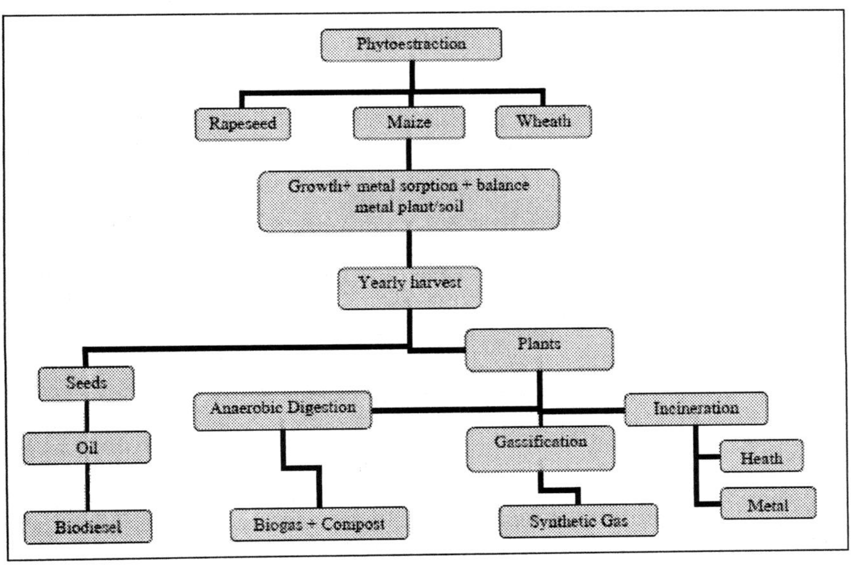

Chapter 4

APPLICATIONS

The assessment of phytoremediation suitability to remediate contaminated sites is in progress since the '80s, and allowed identification of the fundamental items for its application, namely:

- laboratory research aimed at a better knowledge of processes regulating metal sorption by plants;
- field and laboratory research for new plant species with high metal sorption capacity, also with genetic engineering techniques;
- laboratory trials and pilot scale experiments to find out the best conditions for application of different phytoremediation technologies;
- full scale experiments, to assess the concrete possibility of applying such technologies to contaminated soils remediation at environmentally and economically sustainable rates.

Metal bioavailability is the metal availability towards a specific living organism (plant, animal, man) in specific environmental conditions (Adriano et al., 1995); therefore, the soil characteristics and the plant (living organism) behaviour control the actual metal availability. The mechanisms involved in such control are not fully elucidated yet, and often, in bioavailability evaluation, the physiological factors, including metal transport through cell membranes, are neglected. Generally, metal concentration is much more elevated in roots than in the aerial parts, but in some instances the opposite happens. Metal accumulation occurs as a consequence of compartmentalization and vacuole complexation, as it was observed in vacuoles isolated from protoplasts of tobacco cells accumulating elevated levels of Cd and Zn (Barbafieri, 2001).

Assimilation of metals, both essential and critical or toxic, is a typical and diffused plant feature (Streit and Strumm, 1993). According to Baker (1981), plants may be subdivided into three categories:

excluder plants: these species may limit metal sorption or translocation to the aerial parts, irrespective of the metal concentration in the substrate, until toxicity symptoms appear. In many cases, the metal is concentrated in roots that create a barrier-effect against translocation (Bini et al., 2008c);

indicator plants: in these species, metal concentration in aerial tissues is proportional to that in soil (Zupan et al., 1995);

accumulator plants: these species may absorb and actively concentrate metals in the aerial parts at levels higher than in soil (Mc Grath and Dunham, 1997).

Among the plants that tolerate high metal concentration, those which present a natural tendency to accumulate very high metal concentrations (up to 500 times more than normal concentration) are considered hyperaccumulator plants (Baker and Brooks, 1989; Baker et al., 1994). More than 400 species are presently known as accumulator plants, and a large quantity are indicator plants, useful in environmental quality evaluation. In current literature, information on new indicator/accumulator plants, in particular species with high biomass, like *Fragmites australis* (Massacci et al., 2001) and *Cannabis sativa* (Kos and Lestan, 2004) is given almost daily, thus contributing to enhance the green technology potentiality.

A provisional list of accumulator plants, with indication of the metal(s) accumulated, is reported in Table 6.

4.1. Phytoextraction of Zinc and Cadmium

Zinc and Cd are chemical elements that present high affinity; it is argued, therefore, that they could be accumulated by the same vegetal species, although with different mechanisms, considering that Zn is an essential micronutrient, while Cd is toxic to plants and animals (Brooks, 1998). Many plants are known as capable to accumulate Zn; most of them belong to the genus *Thlaspi*, with *T. calaminare* accumulating Zn up to 3.96% dry weight. Also *Cardaminopsis halleri*, *Viola calaminare*, *Haumaniastrum katangense* present Zn concentration higher than 10,000 mg/kg, while *Thlaspi caerulescens* only is considered a Cd hyperaccumulator plant, since it may accumulate over 100 mg/kg dry weight (McGrath, 1998). Papoyan and Kochian (2004) studied the metal sorption mechanisms in *Thlaspi*, and concluded that different genes contribute to the hyperaccumulation trait, since they govern processes that can increase the solubility of metals in the root zone, and the transport proteins that move metals into root cells, and from there to the aerial parts of the plant.

Table 6. Provisional list of the main metal accumulator/hyperaccumulator plants

Plant species	Metal accumulated	Reference
Aeollanthus biformifolius	Co, Cu	8
Agrostis capillaris, A. tenuis	Pb, Cr	7,8,12
Altermathera sessilis	Al	1
Alyssum bertoloni, A.murale	Ni	2,3,14
Arenaria patula	Zn	8
Armeria maritima	Cu	2,7
Astragalus sp.	Se	2,8,11
Atriplex sp.	Se	8,11
Becium homblei	Cu	2,7
Berkeyia coddii	Ni	8
Bonmullera sp.	Ni	8
Brassica jucea, B. napus	Pb,Se	11
Buxus sp.	Ni	8
Calendula arvensis, C. officinalis	Cr	15
Cardaminopsis halleri	Zn	8,11
Eichornia crassipes	Cr, As, Cd, Hg, Pb	1,8
Elodea canadensis	As, Co, Cu, Ni	8
Equisetum arvense	Cu, Zn	8
Festuca arundinacea, F. ovina	Se, Cu, Zn, Cr, Ni, Pb	7,11,13
Haumaniastrum robertii	Co, Cu	8
Hordeum vulgaris	Hg	5
Hybanthus floribundus	Ni	8
Ipomea alpina, I. carnea	Cr, Zn, Cu, Pb, Cd	1,8
Lemna minor	Cr, Cd, Cu, Ni, Pb, Se	1,12
Lolium perenne	Cu, Zn, Cr, Ni, Pb	13
Minuartia verna	Cu, Co	2,7,8
Myriophyllum verticillatum	As, Co, Cu, Mn	8
Phyllantus sp.	Ni	8
Picea abies	U, Hg	4,9
Pinus Nigra, P. silvestris	U, Hg (Pb, Zn)	4,9,10
Plantago lanceolata	Cu, Cd, Pb, Zn	16

Table 6. (Continued).

Plant species	Metal accumulated	Reference
Populus nigra	Pb, Zn, Cu, Cr, Cd	4
Potamogeton ricardsonii	Co, Cu, Pb, Zn	8
Raphanus sativus	Cd, Zn	14
Salix viminalis	Cd, Zn	14
Sambucus nigra	U	4,9
Saxifraga sp.	Ni	8
Senecio coronatus	Ni	8
Silene vulgaris, S. cobalticola	Cu, Co	2,7
Taraxacum officinale	Cu, Cd, Pb, Cr	17
Thlapsi alpestre	Cu, Co, Zn	8
Thlapsi caerulescens	Zn, Cd	3,8,11,14
Thlaspi calaminare	Zn	3,8
Thlaspi rotundifolium	Zn, Pb	7
Typha angustata	Cr	1
Trifolium pratense	Cd, Zn	6
Viola calaminaria	Zn	3,8
Viscaria alpina	Cu	2
Zea mays	Cd, Zn	14

1 - Mhatre and Pankhurst (1997);
2 – Pandolfini et al. (1997);
3 – Baker and Brooks (1989);
4 – Wagner (1993);
5 – Panda et al. (1992);
6 – Kabata-Pendias et al. (1993);
7 – Ernst (1996);
8 – Brooks (1998);
9 – Steubing and Haneke (1993);
10 – Bargagli (1993);
11 – Mc Grath (1998);
12 – Zayed et al. (1998);
13 – Pichtel and Salt (1998);
14 – Felix (1998) ;
15 – Bini et al. (2000a);
16 – Zupan et al (1995);
17 - Bini et al. (2000b)

Other plants, like *Brassica juncea*, proved effective to remove Cd from the soil, although not at hyperaccumulator level (Kumar et al., 1995). Many other plants, however, are likely to accumulate Cd; better knowledge and more research would enhance identification of new Cd accumulator species.

Thlaspi caerulescens requires a minimum concentration of micronutrient Zn much higher than the mean Zn concentration in non-accumulator plants, presumably 1,000 mg/kg in the soil (McGrath, 1998). Mean Zn concentration in tissues of *Thlaspi caerulescens* is much higher than in "normal" plants (up to 30,000 mg/kg vs 100 mg/kg), while exhibiting few or no toxicity symptoms. It is likely the mechanisms responsible for the high metal tolerance to be effective even at low metal concentration in the substrate; this may impede or limit the casual diffusion of this species in the surroundings of the contaminated area, assuring in the meantime adequate Zn level for growing "normal" plants.

4.2. Phytoextraction of Lead

Lead phytoextraction is difficult because of the strong binding of the metal to the organic matter and the soil mineral fraction, which limits translocation to the aerial parts. At present, moreover, a few species are known to be able to accumulate Pb; among them, *Thlaspi rotundifolium* may accumulate up to 8,200 mg/kg d.w.. High Pb levels in plant tissues could also derive from foliar adsorption of aerial input, a common way related to vehicular traffic until the '90s; this Pb source, of course, should not be accounted for in terms of phytoextraction.

To attain effective decontamination in a relatively short time, plants having Pb concentration up to 10,000mg/kg, and with a biomass higher than 20t/ha/y (dry weight) should be necessary (McGrath, 1998). The highest Pb levels (up to 3.5% d.w.) have been found by Kumar et al. (1995) in an experimental trial on different species of Brassicaceae, with *Brassica juncea* producing 18t/ha biomass. These results suggest that Brassicaceae are likely to remove Pb from contaminated soils at a rate of 630kg/ha/y. Phytoextraction coefficients at full scale level, however, are less likely than those obtained with controlled conditions, as shown by Huang and Cunningham (1996), and reported in Table 7.

The most promising method of Pb phytoextracting is to utilize syntetic chelatants to enhance metal assimilation by plants (assisted phytoextraction). Huang and Cunningham (1996) experimented different chelatants on several plants, and in their experiment EDTA proved the most effective. Phytoextraction coefficients up to 1000 times higher than in absence of chelatants were assessed, also with native non-accumulator plants. In such case, plants present early toxic symptoms, and therefore they should be harvested before senescence or death. According to the same Authors, two harvests of *Zea mays* yielding 25 t/ha/y , with Pb concentration up to 10,500 mg/kg, would decrease Pb level in soil to 600 mg/kg in only seven years.

Table 7. Lead concentration (µg/g) in shoots of various species, both native and cultivated (Huang and Cunningham, 1996)

Species (cultivar)	Hydroponic Culture	Soil Culture
Zea mays	375	225
Brassica juncea (211000)	347	129
Brassica juncea (531268)	241	97
Thlaspi rutundifolium	226	79
Triticum aestivum	139	12
Ambrosia artemisiifolia	96	75
Brassica juncea Cern.	65	45
Thlaspi caerulenses	64	58

In case of chelatant application, the decontamination project should pay attention to percolating waters, in order to avoid groundwater contamination. Moreover, foliar fertilization with phosphate would be necessary to preserve nutritional levels.

4. 3. Phytoextraction of Copper and Cobalt

A few examples of plants accumulating copper are known at present, all coming from Cu-Co mine areas of Zaire (Brooks, 1998). In the mineralized area, indeed, many endemic species, characterized by very high tolerance towards these two metals, do exist, and accumulate up to 1% of the two metals. Twentysix plants are Co-hyperccumulator, and 24 are Cu-hyperaccumulator; of these, 9 hyperaccumulate both Co and Cu (Table 8).

The highest Cu concentration in a phanerogame (1.37% d.w.) has been recorded in *Aeollanthus biformifolius* and in *A. subacaulis* var. *linearis*. However, evidence for a possible utilization of such plants in soil cleaning up is lacking (McGrath, 1998). Brooks and Robinson (1998) carried out a study on the opportunity to utilize hyperaccumulator plants as phytomining, and estimated a recovery potential of 0,015 t/ha Cu and Co, with *H. katangense*, assuming to achieve a biomass of 7,5 t/ha after fertilization.

Table 8. Vegetal species hyperaccumulating Cu and Co (µg/g) (Brooks 1998)

Species	Cu	Co
Aeollanthus biformifolius	3920	2820
Anisopappus davyi	2889	2650
Buchnera henriquesii	3520	2435
Bulbostylis mucronata	7783	2130
Gutembergia cupricola	5095	2309
Haumaniastrum katangense	8356	2240
Haumaniastrum robertii	2070	10200
Lindernia perennis	9322	2300
Pandiaka metallorum	6260	2139

4.4. PHYTOEXTRACTION OF RADIONUCLIDES

Radionuclide ^{137}Cs contamination has become a major concern in eastern Europe after the Chernobyl accident (1986). Previously, the cause of ^{137}Cs contamination was the aboveground nuclear testing. Although this poisonous activity has been drastically reduced, large land areas are still polluted with radiocesium. Cesium is a long-lived radioisotope with a half-life of 32.2 years. Because of its low mobility, it contaminates the first few centimeters (topsoil). The primary limitation to removing cesium from soils with plants is its (low) bioavailability (Bini et al., 1991). Indeed, the form of the element makes it unavailable to the plants for uptake.

Recent works by Fuhrmann and co-workers (Fuhrmann et al., 2003) found the ammonium ion was most effective in dissolving ^{137}Cs in soils. This treatment increased the availability of cesium-137 for root uptake and significantly stimulated radioactive cesium accumulation in plant shoots by chemically assisted phytoestraction.

Field studies carried out with three different plant species at the contaminated site in Brookhaven National Laboratory, New York, showed significant variations in the effectiveness of plant species for cleaning up Cs-contaminated sites. *Amaranthus retroflexus* was up to 40 times more effective than other plants in removing 3 percent of the total radiocesium from soil, in just one 3-month growing season, proving that, with two or three yearly crops, the plant could clean up the contaminated site in less than 15 years (Fuhrmann et al., 2003). After harvesting crops, the plant biomass may be disposed in landfill, with or without ashing

The same technology proved to work with uranium. Uranium contamination has become a major concern for human health after the II World War, and particularly during the NATO attack to Bosnia, in the early '90s, and later in Iraq and Afganistan. For soil contaminated with uranium, Kochian (source: http://www.ars.usda.gov/is/AR/archive/jun00/soil0600.htm) found that adding the organic acid citrate to soils greatly increases both the solubility of uranium and its bioavailability for plant uptake and translocation. Citrate operates by binding to insoluble uranium in the soil.

With the citrate treatment, shoots of test plants increased their uranium concentration to over 2,000 mg/kg, 100 times higher than the control plants. This demonstrates the possibility of using citrate -an inexpensive soil amendment - to help plants reduce uranium contamination.

The projected cost of cleaning up these radionuclide-contaminated soils, however, is still very high, exceeding 300 billion dollars.

4. 5. Phytotoxicity Symptoms/Tolerance

Environmental contamination with heavy metals is more and more increasing due to waste materials in the form of compost, sludge and dry residues intensively used as fertilizers and as soil additives in agriculture, and may cause phytotoxic effects and severe alteration in plant structure and function, such as a reduced development of the whole plant. Furthermore, bioaccumulation may occur in plant tissues, leading to further translocation of harmful elements to the food chain. This increases the toxic effect hazard to both humans and animals (Bini et al., 2000a).

Nevertheless, plants exhibit a much higher tolerance to poisoning with these substances than animals (Corradi et al., 1993). This is particularly true of plants growing on naturally contaminated soils (e.g. serpentine soils), since they are adapted to these particular ecological conditions, i.e. they are genetically tolerant, and therefore of interest to phytoremediation.

As previously stated, in the presence of heavy metals, plants present phytotoxicity symptoms, manifested trough a reduced development of the whole plant (roots, stem, leaves). Studies carried out on the toxic effects of Cr in different plants (marigold, sage, dandelion, thime) have shown the following results.

In marigold wild specimens (*Calendula arvensis*, *C. officinalis*) treated with different Cr(III) concentration solutions, Maleci et al. (2001) found a reduction of the meristematic zone of the root tip, and early tissue differentiation (Figure 4a,b), and therefore a reduced elongation of the roots.

In wild sage (*Salvia sclarea* L.) grown in pot and treated with selected Cr(VI) concentrations, Corradi et al. (1993) noted that, although seed germination was not affected, when the emergent radicle came in contact with the Cr solution, its growth was inhibited, although early shoots and cotyledons developed normally: Moreover, reduction in root size, damaged root cap and epidermal cells, collapsed trichomes and root hairs, chlorosis and depressed carotenoid content were observed.

In dandelion (*Taraxacum officinale*) cultivated in pot with compost, and also in wild specimens grown on Cr-contaminated soil, Bini et al. (2000b; 2008b) found that metal uptake and translocation to the aerial parts was reduced, except for Zn, which has an antagonist effect with other metals. This is particularly evident with chromium, which accumulates in roots (barrier effect) (Figure 5a,b). However, no toxic symptoms were observed in both experiments, suggesting dandelion to be a tolerant/excluder plant.

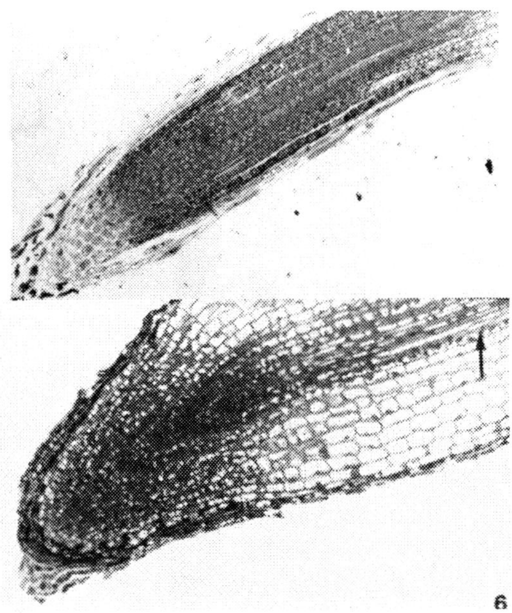

Figure 4. Differential development of the root tip of wild marigold (*Calendula arvensis*) stained with *Toluidine Blu* (light microscopy). a)control seedling (not treated); b) seedling treated with Cr^{VI} (1 mg/kg). Arrow indicates the deformed woody vessels (xylem). Reduction of the meristematic zone and early tissue differentiation are evident. (courtesy L. Maleci)

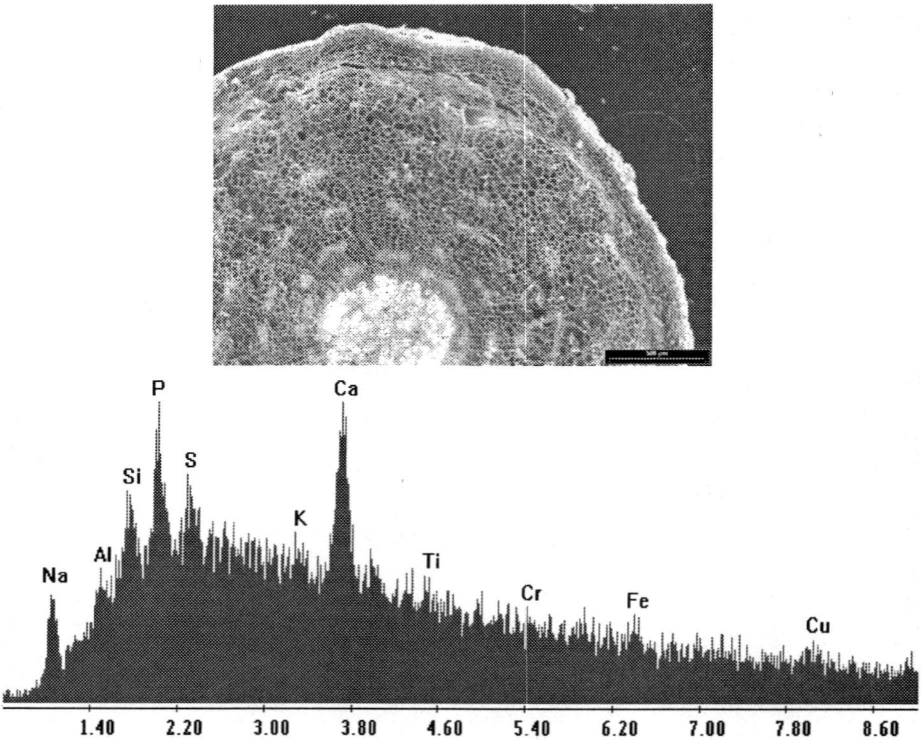

Figure 5 – SEM-EDX observations of wild dandelion (Taraxacum officinale) growing on a tannery-contaminated soil. a) root cross section showing differences in tissue development; b)elemental spectrum of the above root section. Chromium is accounted for 2.3 % d. w. in xylem, and 0.93% d.w. in leaves, suggesting a barrier effect (source: Bini et al., 2008b).

Also plants growing on naturally contaminated soils (e.g. *Alyssum bertoloni*, *Alyssum murale*, *Thymus striatus* ssp. *ophioliticus* on serpentine soils- Figure 6) present particular characters in comparison to plants of the same species, growing on not contaminated soils (Maleci et al., 1999): reduced internodes, highly lignified stem, abundant anthocyanins. It is likely that these plants are affected by the "serpentine syndrome", as defined by Jenny (1989), but they do not present particularly toxic effects, since they are genetically able to tolerate high amounts of heavy metals.

Figure 6. Specimen of wild thime (*Thymus ophioliticus*) fully blossoming on serpentine soil suggests this plant to be tolerant to heavy metals. (courtesy L. Maleci).

The common rock-rose (*Cistus salvifolius* – Figure 7) was selected as a tolerant/indicator plant of contaminated mine sites, since during a phytomining survey (Bini, 2005) it proved to have high biomass, rapid growth rate, high Zn transfer coefficient to leaves (up to 608 mg/kg –Figure 8), high availability of Cu and Pb in roots (127 mg/kg and 111 mg/kg, respectively), and, moreover, no antagonistic neither toxic effects had been observed.

In rock-rose we observed also a different behaviour in the uptake of essential and non essential elements (Bini and Gaballo, 2006). The transfer coefficient is higher for Zn than for Cu and Pb (approx. one order of magnitude: 2.5 vs 0.25). This suggests that the plant is able to regulate the essential Zn translocation from roots to leaves, while translocation of the critical Cu and the phytotoxic Pb is slowed down or even arrested by a root-barrier effect (Figure 9). Therefore, *Cistus salvifolius* can be considered suitable for phytoextraction of Zn and for phytostabilization of Cu and Pb.

Figure 7. Specimens of wild rock-rose (*Cistus salvifolius*) flowering on abandoned mine soil in Mediterranean environment, Italy. (courtesy S. Gaballo).

Aluminum, which is the third most abundant element in the Earth's crust, and is a major component of clays in soil, may constitute a concern to plants growing on acid soils, since at low pH values (<4.5) it is solubilized into the soil solution in a form (Al^{3+}, $AlOH^{2+}$) that is quite toxic to plant roots.

For years, scientists have been looking for the causes of aluminum toxicity in plants (e.g. Foy and Brown, 1964; Pavan and Bingham, 1982; Cameron et al., 1986). Acid soils cover well over 40% of the world's 16 billion hectars of otherwise arable land, including about 180 million hectars in the United States (Maron et al., 2008). When soils become acid, the toxic aluminum damages plant root systems, via inhibition of root growth. For instance, Al-phytotoxicity on coffee seedlings was observed by Pavan and Bingham (1982) in the form of limited and irregular root elongation, color changes, leaf necrosis, as a consequence of increased Al, and decreased Ca and P uptake.

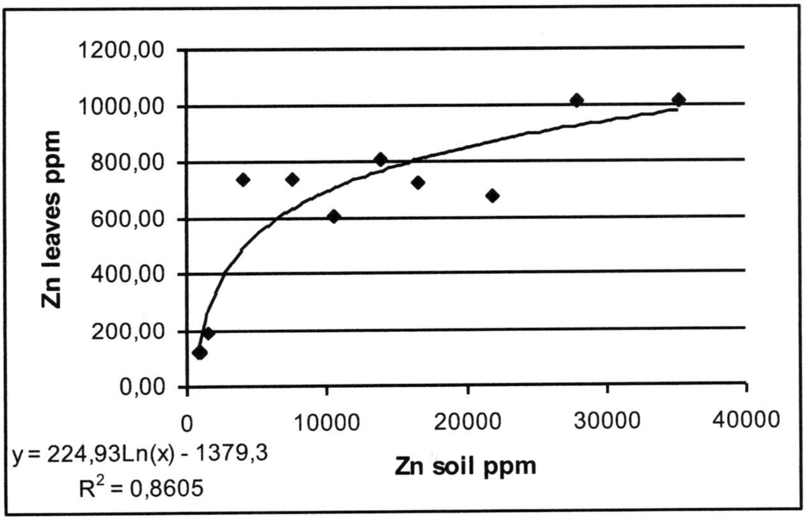

Figure 8. Relationship between zinc concentration in soil and leaves of *Cistus salvifolius*, showing its ability to uptake such essential element as an accumulator plant. (source: Bini, 2005).

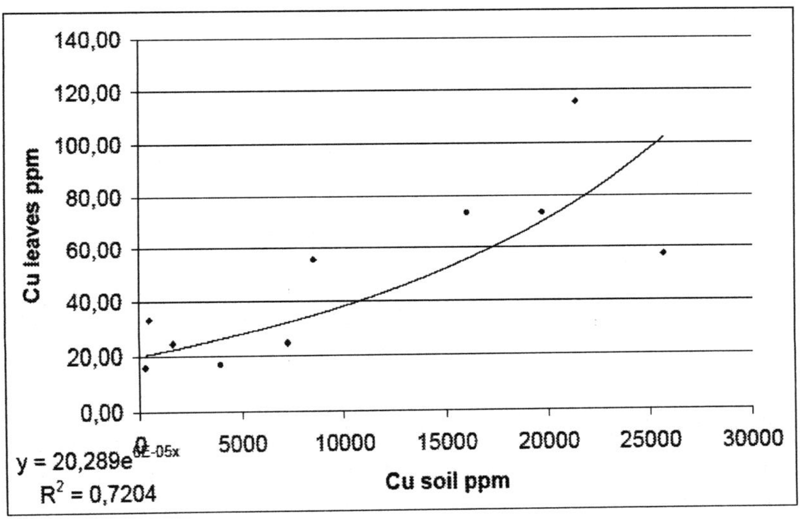

Figure 9. Relationship between copper concentration in soil and leaves of *Cistus salvifolius*, showing its low sorption capacityfor such critical element. (source: Bini, 2005).

The root tip is the key site of injury, leading to a stunted root system, to inhibited root growth, and reduced yields or crop failures from decreased uptake of water and nutrients.

When organic acids are released from the root tip in the soil solution, these acids form a complex with the toxic aluminum, preventing the metal‘s entry into the root. In recent years (see Maron et al, 2008, and references therein), experimental works have been carried out with interdisciplinary approach, integrating molecular, genetic, and physiological research, in order to provide insights into plant tolerance to high levels of Al in soils. Several plants have been tested to highlight Al-toxicity tolerance mechanisms: *Arabidopsis thaliana*, a member of the mustard family, was found to have mutants that are aluminum tolerant. At the U.S.D.A. ARS laboratories (source: http://www.ars.usda.gov/is/AR/archive/jun00/soil0600.htm), researchers are studying differences between these mutants and a wild type of *Arabidopsis* to identify the genetical basis of tolerance, to improve the tolerance of relatively aluminium-sensitive species, and to produce crop species genotypes with increased aluminium tolerance. Wheat and maize proved tolerate aluminum by excluding the metal from the root tip (Maron et al., 2008), thereby giving opportunities for farmers to cultivate marginal acid lands that are not currently used for food production.

Chapter 5

SOME STUDY CASES

Numerous experimental works have been carried out in last years, both in laboratory, in batch and at full scale, aimed at assessing phytoremediation effectiveness, and implementing its potentiality, with the perspective of taking economical advantage, besides the environmental restoration of contaminated land. Blaylock et al. (1997) developed a technology which combines the ability of a Pb-accumulator plant, *Brassica juncea*, with fertilization of a large, densely inhabited district, where several cases of Pb poisoning in children were recorded. After the first cultivation cycle, the most contaminated zone (originally up to 1,000mg/kg Pb), was reduced by 25 %, and Pb concentration was 800 mg/kg.

Similar results are reported by Shen et al. (2001) after EDTA application to a mine soil cultivated with *Brassica napus*. Among herbaceous native plants, *Plantago lanceolata* and *Taraxacum officinale* have been long tested as metal indicator/accumulator plants, in soils with different contamination levels (Zupan et al., 1995, 2003; Bini et al., 2000b; 2007).

Some plants, like mint (*Mentha aquatica*), fern (*Pteris vittata*) and cane (*Phragmites australis*), owing to their high biomass, proved particularly effective in heavy metal rhizofiltration, in particular As, in hydromorphic conditions (wetland areas, submerged soils). Sette et al. (2001) carried out a research project in a wetland area close to an abandoned Sb-As mine, and found that *Mentha aquatica* may accumulate in roots up to 8,900 mg/kg As; instead, translocation to shoots and leaves is very little (13 and 177 mg/kg, respectively). The fern too, in particular *Pteris vittata*, proved effective (Table 9) to accumulate high amounts (up to 4,300 mg/kg) of As (Blaylock e al, 2003; Caille et al., 2003), with high transfer coefficient (As plant/As soil = 9).

Table 9. As accumulation in three species of fern (after Blaylock, 2003)

Species	As concentration mg/kg	Yield kg/ha	Recovery mg/kg
Pteris vittata	900	13050	5.9
Pteris mayii	2013	6100	6.1
Pteris parerii	1416	5050	3.6

Phragmites australis proved effective metal accumulator in numerous experimental trials (Yez et al., 1997; Massacci et al., 2001), besides being currently utilized in domestic and industrial wastewater phytodepuration plants (Figure10), especially for removing nitrate and phosphate. A peculiar character of this plant is the genetic similarity to cereal plants currently cultivated, like maize and wheat, which could be utilized in phytoremediation.

Figure 10. Wild cane (*Phragmites australis*) is currently utilized in phytodepuration (rhizofiltration) of domestic waste, since it proved effective in bioremediation and metal accumulation as well. (photo Bini)

Experimental research on accumulator plants with cultivated species is by far below the expected, although it is more and more increasing in progress. One of the most investigated plants, togheter with maize, is sunflower (*Heliantus annuus*). This plant, grown on a highly metal-contaminated soil and treated with EDTA (assisted phytoextraction), showed noteworthy ability to accumulate metals in its tissues (Quartacci et al., 2001). Thirty days after seedling, it accumulated 2,500 µg Zn, 500 µg Cu, Cd e Pb, 100 µg Cr. Most of metals (over 50%) were concentrated in roots: only Zn proved good metal transfer capacity to aerial parts (50% in shoots, 25% in leaves). Further investigations with sunflower (Sacchi et al., 2001), however, showed important limitations in Pb-assisted phytoextraction: the application of the maximum rate of chelatant determined Pb increases not proportional to the lower rate, and a concentration of 600 mg/kg Pb in leaves. According to the authors, given a biomass yield of 10 t/ha, lead would be removed from the contaminated soil at a rate of 6 kg/ha, in more than 100 years, a not cost-effective time; moreover, chelatant application would be too expensive and pose serious environmental hazard because of the possible leaching of chelatant to subsoil and groundwater.

Arboreal plants, until now, received little attention in comparison to the herbaceous ones, due to their different physiology; however, their high biomass would constitute an important item to their use in phytoremediation. In last years, indeed, experimental research on arboreal plants has progressed quickly, and interesting results have been attained, especially with short-term coppices such as willow and poplar. One of the most experimented plant is willow (*Salix viminalis*, *S. caprea*, *S. rubra*, *S. fragilis*).

Greger e Landberg (2003) recorded a strong correlation between *Salix viminalis* biomass and metal uptake from soil, and Wieshammer et al. (2003) found high metal concentrations (326 mg/kg Cd, 2,413 mg/kg Zn, 70 mg/kg Pb) in willow leaves, with increments up to 54% Zn, 39% Cd, 21% Pb, in mycorrhized plants inoculated with metal-tolerant bacteria.

A survey carried out in an urban park of Great Britain (Lepp e Dickinson, 2003) showed tha 50 years after birch (*Betula* sp.), maple (*Acer pseudoplatanus*) and willow (*Salix caprea*) plantation on a strongly degraded and contaminated area, the vegetation cover had been naturally reconstructed, suggesting the potential resilience of natural vegetation to be effective in remediation of contaminated soils.

A major concern in phytoremediation is that most plants present little biomass, and therefore their ability to take up metals in the aerial parts is quite limited. To overcome this problem, experimental work both in pot (batch scale – Figure 11) and in the field (pilot scale – Figure 12) is in progress since the '80s, in order to find out new accumulator plants, and to increase biomass of just known accumulator species, by genetic engineering techniques.

Figure 11. Pot and batch experiments with marigold (*Calendula officinalis, C. arvensis*) allowed to control the whole plant development (root elongation, shoot growth, metal uptake, etc.) during treatment with hevy metal-bearing solutions. (source: Bini et al., 2000a).

Figure 12 – Pilot scale and full scale field experimental trials with dandelion (*Taraxacum officinale*) permited to assess plant tolerance to different metal-bearing treatments, ant to estimate the fertility gain (increased biomass) at harvesting, thereby suggesting metal recovery. (source: Bini et al., 2000b)

Among the most studied plants (Table 10) are many Brassicaceae, and particularly *Thlaspi caerulescens*, which proved able to uptake high amounts of Zn and Cd (Mc Grath, 1998), and *Brassica juncea*, a Pb accumulator (Marchiol et al., 2004).

Among crop species, maize proved effective in Pb phytoremediation (Huang and Cunningham, 1996), and sunflower seems effective in phytostabilization, showing a root-barrier effect (Simon, 2003).

Table 10. Some examples of accumulator plants, with indication of metal concentration, biomass production and metal recovery

Plant Species	Zn Concentration (kg/t)	Biomass Production (t/ha)	Metal Recovery (kg/ha)
Thlaspi caerulescens	10.63	5.4	57.4
Cardaminopsis halleri	5.21	2.7	14.1
	Pb Concentration (kg/t)	Biomass Production (t/ha)	Metal Recovery (kg/ha)
Zea mays	10.5	25	1900
Brassica juncea	1.29	18	630
	As Concentration (kg/t)	Biomass Production (t/ha)	Metal Recovery (kg/ha)
Pteris vittata	0.9	13	20

Other species have been experimented for As rhizofiltration in aquatic systems. *Mentha aquatica* and *Pteris vittata* proved effective in As accumulation in roots and shoots (up to 8,900 mg/kg and 4300 mg/kg, respectively (Blaylock et al, 2003). Some other plants, such as *Phragmites australis* and *Cannabis sativa*, owing to their high biomass, were experimented in phytodegradation of nitrate and phosphate, although their metal uptake capacity is limited.

Identification of arboreal plants with high biomass and rapid grow-up is the new frontier of phytoremediation technology. These plants should be able to uptake metals from contaminated soils, and translocate them to stems and ultimately to leaves. Afterwards, they could be harvested, and the metal recovered. This is what has been experimented with *Salix viminalis* and *Salix caprea* in Sweden, achieving very good metal recovery: up to 2,413 mg/kg Zn, and 326 mg/kg Cd (Greger and Landberg, 2003).

Chapter 6

CONCLUSION

The restoration of contaminated sites is one of the most important environmental issues. Soil pollution by chemicals poses serious hazards to surface and ground waters, plants and humans, and presents relevant social, sanitary and economic costs. In the U.S.A., for instance, the projected ex-situ soil decontamination cost is up to 3 million $ per hectar, while it is estimated to decrease to 0.25 million $ per hectar by bioremediation.

The assessment of soil contamination has been extensively carried out through plant analysis. The estimation of metal uptake by plants, combined with geobotanical observations, proved an useful tool to find tolerant plant populations to be used in revegetation programs aimed at reducing the environmental impact of contaminated areas. Wild and cultivated plant species have been used as (passive accumulative) bioindicators for large scale and local soil contamination, as reported in the above examples. It is proved also that tolerant or accumulator populations of higher plants may colonize naturally or even anthropogenic metal-enriched areas, accompanying the disappearance of sensitive plants.

Among all the remediation technologies currently utilized, in last years great attention has been paid to phytoremediation, given the numerous advantages it offers (green technology, simple concept, cost-effective) with respect to current engineering-based technologies, which are very costly and dramatically disturb the landscape. Phytoremediation is an emerging technology that holds great potential in cleaning up contaminants that: 1) are near the surface, 2) are relatively non-leachable, 3) pose little imminent risk to human health or the environment, and 4) cover large surface areas. Moreover, it is cost-effective in comparison to current technologies, and environmental friendly.

However, there are two questions. The first is the inadequate understanding of metal transport in plants and tolerance mechanisms. To address this deficit, search of new accumulator plants is needed, in order to elucidate the fundamental mechanisms of metal accumulation, and ultimately the genes that regulate the amount of metals taken up from the soil and deposited at other locations within the plant (stems, shoots and ultimately leaves). Thereby, the choice of plants is a crucial aspect for the remediation techniques.

The second question is that many accumulator plants have little biomass and a slow growth rate. Therefore, there is an urgent need for surveying and screening of plants with ability to accumulate metals in their tissues, and a relatively high biomass, or improving, by genetic engineering, the plant biomass, in order to provide economic metal recovery after harvesting.

Up to now, most experimental trials have been carried out at micro or meso scale (in microcosm or in pot), while only few full scale experiments have been carried out in the field. Nevertheless, such experiments could help in assessing the feasibility of phytoremediation, and its effective contribution to clean-up contaminated soils.

However, phytoremediation is not yet ready for full scale application, despite favourable initial cost projections, which indicate expansion of clean-up market to be likely in next years (Table 11).

Table 11. Phytoremediation market projection for past and future years in U.S. (millions U.S. $) (adapted from Glass et al., 1997)

Site categories	2000	2005	2010
Groundwater polluted with organic compounds	2-3	10-15	20-45
Heavy metal contaminated soils	1-2	15-22	40-80
Radionuclide contaminated soils	0.1-0.5	2- 5	25-30
Waste water polluted with heavy metals	0.1-1.5	1- 3	3- 5
Other sources	0-1	2- 5	12- 20
Whole U.S. phytoremediation market	3-8	30-50	100-180

In conclusion, more effort should be done to improve this technology. Fundamental items for future research in this perspective are:

- to find out new highly efficient accumulator plants, and related cultivation technologies;
- better knowledge of processes and mechanisms regulating metal uptake by plants;

- pilot scale and laboratory trials, to find out the most suitable conditions for applying the different remediation techniques;
- Full scale and field trials, to assess the actual feasibility of these techniques according to criteria of environmental and economic sustainability, in terms of costs and benefits.

Moreover, basic research is necessary to advance our scientific understanding of physical, biological, and chemical processes important for soil remediation. More research is needed in many disciplines such as microbiology, molecular biology, genetic engineering, geochemistry, hydrology, and transport processes. Finally, basic research should be focused on the behaviour of complex systems that include mixtures of contaminants and organisms, and this research must account for the spatial and temporal variability of such systems.

Although more work is expected before possible commercialisation of this technology, it is current opinion, among scientists of the different disciplines involved, that profound insight into phytoestraction could give appreciable results in the application of new intervention techniques for the restoration of contaminated soils at low environmental impact, and low cost, and that phytoremediation could acquire a noteworthy portion of the clean-up market in the next future.

REFERENCES

Adriano D.C., Bollag J.M., Frankemberger W.T., Sims R.C. (1999) – Biodegradation of contaminated soils. Agronomy Monograph 372, S*oil Science Society of America*, pp772.

Adriano D.C., Chlopecka A., Kapland D.I., Clijsters H., Vangrosvelt J. (1995) – Soil contamination and remediation philosophy, science and technology. In: *Contaminated Soils* (R. Prost ed.) INRA, Paris, 466-504.

Adriano, D.C. (2001) - Trace Elements in the Terrestrial Environment.2th edit., Springer-Verlag., New York, N.Y. 866p

Alloway, B.J. (1995) - Heavy Metals in Soil. Chapman and Hall. London, England.

Argese E., Delauney E., Agnoli F., Faraon F., Sorgano A., Cacco G. (2001) - Variazione di parametri morfo-fisiologici di piante di orzo ed avena allevate in microcosmi trattati con metalli pesanti. *Boll. Soc. It. Sci. Suolo*, 50, 3, 709-722.

Baize D. and Sterckeman T. (2004) – On the necessity of knowledge of the natural pedo-geochemical background contents in the evaluation of contamination of soils by trace elements. *Sci. Total Envir.*, 264, 127-139.

Baker A. J. M. (1981) – Accumulators and excluders strategies in the response of plants to heavy metals: *J. of Plant Nutrition,* 3, 643-654.

Baker A. M. J., Brooks R. (1989) – Terrestrial higher plants which hyperaccumulate metallic elements – A review of their distribution, ecology and phytochemistry. *Biorecovery*, 1, 81-126.

Baker A., Mc Grath S., Reeves R., Smith J. (2000) – Metal hyperaccumulator plants: a review of the ecology and physiology of a biological resource for phytoremediation of metal-polluted soils. In Phytoremediation of contaminated soils (N. Terry and G. Banuelos eds), Lewis Publisher, London, pp85-107.

Baker, A., S.P. Mcgrath, C.M.D. Sidoli, And R.D. Reeves. 1994. The possibility of in-situ heavy metal decontamination of polluted soils using crops of metal-accumulating plants. *Resource Conserv. Recycling.* 11:41-91.

Barbafieri M. (2001) – Applicabilità e limiti della fitodepurazione. In: Siti contaminati: indagini, analisi di rischio e tecniche di bonifica (a cura di L. Bonomo). Politecnico di Milano, pp 699-717.

Bargagli R. (1993) – Plant leaves and lichens as biomonitors of naturals or anthropogenic emissions of mercury. In B. Markert Ed. Plants as biomonitors. Weinheim WCH, pp 468 – 484

Bini C. (2005) – Plants growing on abandoned mine soils: A chance in phytoremediation. Proc. III EGU Conf., Wien (CD-rom).

Bini C. (2008a) - Fate of trace elements in the Venice lagoon watershed and conterminous areas (Italy). Novapublisher (in press).

Bini C. (2008b) - Soil: a precious natural resource. Novapublisher (in press).

Bini C., Gaballo S. (2006) – Pedogenic trends in Anthrosols developed in sulfidic mine spoils: A case study in the Temperino mine archaeological area (Campiglia Marittima, Tuscany, Italy). *Quat. Intern.*, 156-157, 70-78.

Bini C., Casaril S., Pavoni B. (2000b) – Fertility gain and heavy metal accumulation in plants and soils. *Tox. Environ. Chem.* 77: 131 – 142

Bini C., Maleci L., Romanin A. (2008c) - The chromium issue in soils of the leather tannery district in Italy. *Jour. of Geoch. Explor.*, 96, 2-3, 194-202

Bini C., Maleci L., Gabbrielli L., Paolillo A. (2000a) – Biological perspectives in soil remediation with reference to chromium . In bioremediation of contaminated soils D.Wise Ed. Marcel Dekker Inc., N.Y., pp.663 – 675

Bini C., Giovani C., Padovani R., Cesco S., Maggioni A., Mondini C. (1991) - Cs137 transfer to forage in Friuli-Venezia Giulia mountain areas. *Proc. VI Conv. Naz. A. I. F. B.*, Genova.

Black H., (1995) – Absorbing possibilites: phytoremediation. Environmental health perspectives, 103,12: 1 - 6

Blaylock M.J., Elles M.P., Nuttal C.Y. Zdimal K.L. Lee C.R. (2003) – Treatment of As contaminated soil and water using Pteris vittata. Proc VI ICOBTE, Uppsala, Sv.

Blaylock M.J., Salt D.E., Dushenkov S., Zakharova O., Gussman C., Kapulnok Y., Ensley B.D., Raskin I. (1997) – Enhanced accumulation of Pb in Indian Mustard by soil-applied chelating agents. *Environ. Sci and Technol.,* 31, 3, 860-865.

Brooks R.R. (1998) – Phytochemistry of hyperaccumulator. In: Plants that hyperaccumulate heavy metals: their role in phytoremediation, microbiology, archaeology, mineral exploration and phytomining (R. Brooks ed.). CAB International, U.K., 15-53.

Brooks R.R., Robinson B.H. (1998) – The potential use of hyperaccumulators and other plants for phytomining. In: Plants that hyperaccumulate heavy metals: their role in phytoremediation, microbiology, archaeology, mineral exploration and phytomining (R. Brooks ed.). CAB International, U.K., 327-356.

Bryda L.K., Sellman C.L. (1994) – Recent developments in cleanup technology. Remediation (Autumn), 475-489.

Caille N. Swanwick S. Zhao F.J., Mc Grath S. (2003) – Assessment of As phytoestraction potential of two ferns by field and pot experiments. Proc VII ICOBTE , Uppsala, Sv.

Cameron R.S., Ritchie G.S.P., Robson A.D. (1986) – Relative toxicities of inorganic aluminum complexes to barley. *Soil Sci. Soc. Am. J.*, 50, 1231-1236.

Carratù G., Carafa A.M., Aprile G.G. (2001) – Valutazione della capacità di Pb-decontaminante di due specie ornamentali. *Boll. Soc. It. Sci. Suolo*, 50, 3, 733-738.

Chaney R.L., Li Ym., Angle J.S., Baker A. (1997) – Converting metal hyperaccumulator wild plants into commercial phytoremediation systems: approaches and progress. Proc IV ICOBTE, Berkeley, Cal 623 – 624.

Corradi M., Bianchi A., Albasini A. (1993) – Chromium toxicity in *Salvia sclarea*. Effects of hexavalent chromium on seed germination and seedling *develpment. Envir. Exper.* Botany, 33, 3, 405-413.

Cunningham S.D., and Berti W.R. (1997) – Phytoestraction or in-place inactivation: technical, economic and regulatory consideration on the soil-lead issue. Proc IV ICOBTE Berkeley Cal. 627 – 628.

Cunningham, S. D., And W. R. Berti (1993). Remediation of contaminated soils with green plants: An overview. *In Vitro Cell. Dev. Bio.*, 29P:207-212.

Dushenkov, V., P. Dumar, H. Motto And I. Raskin. 1995. Rhizofiltration - The use of plants to remove heavy metals from aqueous streams. *Environ. Sci. Technol.* 29:1239-1245.

Dutch Ministry Of Housing, Spatial Planning And Environment. 1995. Soil protection act. (Parliamentary Paper I, 1993/94, 21556, No. 266. The Hague, Netherlands, 19p.

Eikmann T., Kloke A. (1993) - Ableitungskriterien fur die Eikmann-Kloke-Werte. In: Kreysa, G., and Wiesner, J. (ed): Beurteilung von Schwermetallen in Boden von Ballungsgebieten: Arsen, Blei und Cadmium. Internationale Expertenbeitrage und Resumee der DECHEMA Abeitsgruppe, 469-500.

Eikmann T., Kloke A. (1995) – Nutzungs- und Schutzgutbezogene Orienterungswerte fur (Schad) Stoffe in Boden. In: Rosenkranz D., Einsele G., Harress H.M. (ed): Bodenschtz. Erganzbares Handbuchder Massnahmen und Empefehlungen fur Schutz, Pflege und Sanierung von Boden, Landschaft und Grundwasser, 1: 1-7.

Ernst W.H.O. (1996) – Bioavailability of heavy metals and decontamination of soils by plants. *Applied Geochemistry*, 11, 163-167.

Exner J.H. (1995) – Alternatives to incineration in remediation of soils and sediments assessed. *Remediation*, 5: 1-18.

Felix H. (1998) – Field trials for in-situ decontamination of heavy metal polluted soils using crops of metal accumulating plants. Z.Pflanzernhar. Bodenk, 160: 525-529

Ferguson J.E. (1992) – The heavy elements. Chemistry, environmental impact and health effects. Pergamon Press, pp 383-397.

Foy C.D. and Brown J.C. (1964) – Toxic factors in acid soils. Diffrential Aluminum tolerance of plant species. *Soil Sci. Soc. Am. Proc.*, 27, 403-407.

Fuhrmann, M., Lasat, M.M., Schwartz, M., Ebbs, S.D., Kochian, L.V., Cornish, J. (2003) - Uptake of 137Cs and 90Sr from contaminated soil by three plant species. Application to phytoremediation. *Journal of Environmental Quality*.

Glass D. and Associated, Inc. Estimates (1997) – Estimated U.S. markets for phytoremediation 1997-2005. Explorer-Internet.

Grasso, D. 1993. Hazardous waste site remediation: Source Control. Lewis Publishers. CRC Press Inc., Boca Raton, FL.

Greger M. and Landberg T. (2003) – Improoving removal of metals from soil by willow (Salix viminalis). Proc VII ICOBTE , Uppsala, Sv.

Huang J.W., Cunningham S.D. (1996) – Lead phytoextraction: species variation in lead uptake and translocation. New Phytol., 134, 75-84.

Jenny H. (1941)- Factors of soil formation. McGrow-Hill, N.Y.

Jenny H. (1989) – The soil resource. Springer Verlag, New York.

Kabata-Pendias A. (1997) - Trace metal balance in soil - a current problem in agriculture. In D.C. Adriano, Z. S. Chen and S.S. Yang (eds.). Biogeochemistry of Trace Elements. *Adv. Environ. Sci. Applied Science* Publishers, Northwood, England.

Kabata-Pendias A., Piotrowska M., Dudka S. (1993) – Trace metals in legumes and monocotyledons and their suitability for the assessment of soil contamination. In B. Markert Ed. Plants as biomonitors. Weinheim WCH, pp 485 - 494

Kloke A., Sauerbeck D., Vetter H. (1980) – The contamination of plants and soils with heavy metals and the transport of metals in terrestrial food chains. In:

Nriagu J.O. (ed), Changing metal cycles and human health. Life Science Res. Rep., 28: 113-141. Springer, Berlin.

Kos B. and Lestan D. (2004) - Soil washing of Pb, Zn, and Cd using biodegradable chelator and permeable barriers and induced phytoestraction by Cannabis sativa. *Plant and Soil* 63 1- 2 : 43 -51.

Kramer U. (1996) – Free histidine as a metal chelator in plants that accumulate nickel. *Nature,* 379, 635-638.

Kumar, P., V. Dushenkov, H. Motto, And I. Raskin. 1995. Phytoextration - The use of plants to remove heavy metals from soils. *Environ. Sci. Technol.* 29:1232-1238.

Lasat M.M. (2002) – Phytoestraction of toxic metals: a review of biological mechanisms. *J. Environ: Qual.,* 31, 109-120.

Lee C.C., Huffman G.L., Sasseville S.M. (1990) – Incinerability ranking systems for RCRA hazardous constituents. Haz. WasteHaz. Materials, 7: 385-394.

Lee J.S., Lee S.W., Chon H.T, Kim K.W. (2008) – Evaluation of human exposure to arsenic due to rice ingestion in the vicinity of abandoned Myungbong Au-Ag mine site, Korea. *J. Geochem. Explor.*, 96, 231-235.

Lepp N.W. and Dickson N.M. (2003) – Natural bioremediation of metal polluted soils. A case history from the U.K. Proc VII ICOBTE , Uppsala, Sv.

Li H., Sheng G., Sheng W., Xu O. (2002) – Uptake of trifluralin and lindane from water by ryegrass. *Chemosphere*, 48, 335-341.

Lim H.S., Lee J.S., Chon H.T., Sager M. (2008) – Heavy metal contamination and health risk assessment in the vicinity of the abandoned Songcheon Au-Ag mine in Korea. *J. Geochem. Explor.*, 96, 223-230.

Losi, M.E., C. Amrhein, And W. T. Frankenberger, Jr. 1994. Bioremediation of chromate-contaminated groundwater by reduction and precipitation in surface soils. *J. Environ. Qual.* 23:1141-1150.

Ma L.Q., Komar K.M., Tu C., Zhang W., Cai Y., Kennelly E.D. (2001) – A fern that hyperaccumu lates arsenic. *Nature*, 409, 579.

Maleci L., Gentili L., Pinetti A., Bellesia F., Servettaz O. (1999) – Morphological and phytochemical characters of *Thymus striatus* Vahl growing in Italy. Plant Biosyst., 133,2, 137-144.

Maleci L., Bini C., Paolillo A. (2001) – Chromium (III) uptake by *Calendula arvensis* L. and related phytotoxicity. Proc. VI ICOBTE, Guelph, On., 384 (abstract).

Marchiol L., Sacco P., Assolari S., Zerbi G. (2004) – Reclamation of polluted soils: phytoremediation potential of crop-related BRASSICA species. Water, *Air and Soil Pollution*, 158 (1), 345-356.

Maron, L.G., Matias, K., Mao, C., Menossi, M., Kochian, L.V. (2008) - Transcriptional profiling of Al toxicity and tolerance responses in maize roots. *New Phytologist*. 179:116-128.

Massacci A., Iannelli M.A , Pietrini F. (2001) – Il fitorimedio: organismi vegetali come potenziali agenti disinquinanti. *Bol. Soc. It. Sci Suolo* 50,3: 581 – 588.

Mc Grath S. (1995) – Behaviour of trace elements in terrestrial ecosystems. In Contaminated Soils (R.Prost ed.),. INRA, Paris, 35-54.

Mc Grath S. (1998) – Phytoextraction for soil remediation. In: Plants that hyperaccumulate heavy metals: their role in phytoremediation, microbiology, archaeology, mineral exploration and phytomining (R. Brooks ed.). *CAB International*, U.K., pp261-287.

Mc Grath S. and Dunham S.J. (1997) – Potential phytoextraction of zinc and cadmium from soil using hyperaccumulator plants. Proc. IV ICOBTE, Berkeley, Cal., 625-626.

Mcnair M.R. (1997) – The genetics of metal tolerance and accumulation by plants. Proc. IV ICOBTE , Berkeley, Cal., 647 – 648.

Mehmannavaz R., Prasher S.O., Ahmad D. (2001) – Rhizospheric effects of alfalfa on biotransformation of polychlorinated biphenyls in a contaminated soil augmented with Sinorhizobium meliloti. Process Biochemistry.

Mhatre G.N., Pankhurst C.E. (1997) – Bioindicators to detect contamination of soils with special reference to heavy metals. In Pankhrust C.E. et al. Biological Indicators of soil health. CAB Int., Wallingford, U.K. pp 349- 369.

National Research Council (NRC) (1993) – Soil and water quality: an agenda for agriculture. National Academy Press, Washington D.C., 516p.

Panda K.K., Lenka M., Panda B.B. (1992) – Monitoring and assesment of mercury pollution in the vicinity of a chloralkali plant. *Environ. Pollut.* 76: 33 – 42.

Pandolfini T., Gremigni P., Gabbrielli R. (1997) – Biomonitoring of soil healt by plants. In Pankhrust C.E. et al. Biological Indicators of soil health. CAB Int., Wallingford, U.K. pp 325 - 348

Papoyan, A., Kochian, L.V. 2004. Identification of thlaspi caerulescens genes that may be involved in heavy metal hyperaccumulation and tolerance: characterization of a novel heavy metal transporting atpase. *Plant Physiology*. 136:3814-3823.

Pavan M.A. and Bingham F.T. (1982) – Toxicity of Aluminum to coffee seedlings grown in nutrient solution. *Soil Sci. Am. Proc.,* 46, 14-17.

Pichtel J. and Salt C.A. (1998) Vegetative growth and trace metal accumulation on metalliferous wastes. *J. Environ. Qual.* 27: 618 – 624.

Pierzinsky G. (2003) – Trace element chemistry, contamination and ecotoxicity. In: Gobran G. and Lepp N. (Eds): Biogeochemistry of Trace Elements. Proc. VII ICOBTE, 1:1, 14-15, Uppsala.

Prasad M.N. (2003) – Phytoestraction of metals: role of natural hyperaccumulators, transgenic plants and soil amending chelators. Proc VI ICOBTE, Uppsala, Sv.

Quartacci M.F., Sgherri C.L.M., Navari-Izzo F. (2001) – Fitoestrazione da un suolo contaminato da più metalli: accumulo e tolleranza. *Bol. Soc. It . Sci. Suolo* 50, 3: 649 -660.

Raskin I. (1996) – Plant genetic engineering may help with environmental cleanup (commentary). *Proc. Natl. Acad. Sci. USA*, 93, 3164-3166.

Rebedea, I., R. Edwards, N.W. Lepp, A.J. Lovell (1995) - An investigation into the use of synthetic zeolites for in situ land reclamation. Third International Conference of the Biogeochemistry of Trace Elements. Book of abstracts. Paris.

Reiman C. and Garret R.G. (2005) – Geochemical background – Concept and Reality. *Sci. Tot. Envir.*, 350, 12-27.

Sacchi G.A., Rivetta A., Cocucci M. (1999) – Assorbimento radicale e bio acuumulo di metalli pesanti nelle piante: problematiche e prospettive. In Impatto ambientale di metalli pesanti ed elementi in traccie. Pitagora editrice Bologna, 65 – 76.

Salomons, W. 1995. Environmental impact of metals derived from mining activities: processes, predictions, prevention. *J. Geochem. Explor*. 52: 5-23.

Salt, D. E., M. Blaylock, N. Kumar, V. Dushenkov, B. D. Ensley, I. Chet, And I. Raskin. 1995. Phytoremediation: A novel strategy for the removal of toxic metals from the environment using plants. *Biotechnology*. 13:468-474.

Schnoor J.L. (1995) – Phytoremediation of organic and nutrient contaminants. Environ. Sci. and Technol., 29, n.7.

Sette B., Boscagli A., Riccobono F. (2001) – Mentha aquatica L. as an arsenic accumulator. Proc 6th ICOBTE, Guelph, Ont, Ca, p. 395.

Shen Z.G., Li X.D., Chen H.M., Wang C.C., Chua H. (2001) – Phytoextraction of lead from a contaminated soil using high biomass species of plants. Proc. 6th ICOBTE, Guelph, Ont, Ca, p. 133.

Simon L. (2003) – Cadmium rhizofiltration capacity of Helianthus annus and Brassica juncea. In: Gobran G. and Lepp N. (Eds): Biogeochemistry of Trace Elements. Proc. VII ICOBTE, 1:2, 206-207, Uppsala.

Smith, R. A. H., And A. D. Bradshaw. 1979. The use of metal tolerant plant population for the reclamation of metalliferous wastes. *J. Applied Ecology*. 16:595-612.

Souvent P. and Pirk S. (2001) – Pollution caused by metallic fragments introduced into soil because of World War I activities. *Environmental Geology*, 40, 317-323.

Steubing L., Haneke J. (1993) – Higher plants as indicators of uranium occurrence in soil. In: Markert B. (ed): Plants as biomonitors. Weinheim, VCH, pp155-165.

Streit B., Stumm W. (1993) – Chemical properties of metals and the process of bioaccumulation in terrestrial plants. *In Plants as Biomonitors* (B. Markert ed.), VCH, Weinheim, 31-62.

Stringer, D.A. 1990. Hazard assessment of chemical contaminants in soil. ECETOC Technical. Rep. No. 40; Aneme Louise 250, Brussels, Belgium.

Tobias F.J., Bech J., Sanchez P. (1997) – Statistical approach to discriminate background and anthropogenic input of trace elements in soils of Catalonia, Spain. Water, Air and Soil Poll., 100, 63-78.

Tu S., Ma L., Luongo T. (2004) – Root exsudates and arsenic accumulation in As hyperaccumulating Pteris vittata and non hyperaccumulating Nephrolepis exaltata. *Plant and Soil* 258,1: 9 – 19.

Ungaro F., Ragazzi F., Cappellin R., Giandon P. (2008) – Arsenic concentration in the soils of the Brenta plain (Northern Italy). Mapping the probability of exceeding contamination thresholds. *J. Geochem. Explor.*, 96, 2-3, 117-131.

Vangronsveld, J., F. Van Assche And H. Clijsters. 1995b. Reclamation of a bare industrial area contaminated by non-ferrous metals: in situ metal immobilizaation and revegetations. *Environ. Pollut.* 87:51-59.

Vangronsveld, J., J. Colpaert, And K. Van Tichelen 1996. In Press. Reclamation of a bare industrial area contaminated by non-ferrous metals: physico-chemical and biological evaluation of the durability of soil treatment and revegetation. *Environ. Pollut.*

Vangronsveld, J., J. Sterckx, F. Van Assche And H. Clijsters 1995a. Rehabilitation studies on an old non-ferrous waste on old non-ferrous waste dumping ground: effects of revegetation and metal immobilization by beringite. *J. Geochem. Explor.* 52, 221-229.

Vangronsveld, J., Meers E., Dejonghe W., Geurds, M., Diels L., Defoort B., Beeckman E., Smis J. (2007) – Phytoremediation for heavy metal contaminated soils and combined bio-energy production. In: Zu Y., Lepp N., Naidu R. (Eds), Biogeochemistry of trace elements: environmental protection, remediation and human health. Proc. IX ICOBTE, 162-163, Beijing.

Van Meirvenne M., Meklit T., Verstraete S., De Boever M., Tack F. (2008) – Could shelling in the First World War have increased copper concentrations in the soil around Ypres? *Eur. J. Soil Sci*, 59, 372-379.

Vollmer M.K., Gupta S.K., Krebs R. (1995) - New standards on contaminated soils in Switzerland – Comparison with Dutch and German quality criteria. In*: Contaminated Soils (R.* Prost ed.) INRA, Paris, 445-459.

Wagner G. (1993) - Large scale screening of heavy metals burdens in higer plants. In B. Markert Ed. Plants as biomonitors. Weinheim WCH, pp 425 - 434

Wenzel, W. W., H. Sattler And F. Jockwer. (1993) - Metal hyperaccumulator plants: a survey on species to be potentially used for soil remediation. Agronomy Abstracts. p. 52.

Westinghouse Hanford Company (1994) – Soil washing: bench-scale tests on 116-F-4 Plupo Crib soil. WHC-SD-EN-TI-268. Westinghouse Hanford Co., Richland, WA.

Whiting S.N. (2001) – Rhizosphere bacteria mobilize Zn for hyperaccumulation by Thlaspi caerulescens. *Environ. Sci and Technol.*, 35, 3144-3150.

Wieshammer G., Sommer P., Gorfer M., Strauss J., Wenzel W.W. (2003) – Bioavailable contaminant stripping of Cd and Zn using willows. Proc VII ICOBTE , Uppsala, Sv.

Yetz H., Baker A., Wong M.H. Willis A.J. (1997) – Zinc, lead and cadmium tolerance, uptake and accumulation by Fragmites australis. *Ann. Bot.* 80: 363 - 370

Zaccheo P., Crippa L., Gigliotti C. (2001) – Studio dell'efficienza del mais nella phytoremediation di un suolo contaminato da metalli pesanti. *Boll. Soc. It. Sci. Suolo,* 50, 3, 685-692.

Zayed A., Gowthaman S., Terry N. (1998) – Phytoaccumulation of trace elements by wetland plants: duckweed. *J.Environ. Qual.*, 27: 715 -721

Zonta R., Botter M., Cassin D., Pini R., Scattolin M., Zaggia L. (2007) – Sediment chemical contamination of a shallow water area close to the industrial zone of Porto Marghera (Venice Lagoon, Italy). Marine Pollution Bulletin, 55, 529-542.

Zupan M., Hudnik V., Lobnik F., Kadunc V. (1995) – Accumulation of Pb, Cd, Zn from contaminated soil to various plants and evaluation of soil remediation with indicator plant (Plantago lanceolata) In Contaminated Soils (R. Prost ed.) INRA, Paris, 325-335.

Zupan M., Kralj T., Grcman H., Hudnik V., Lobnik F. (2003) – The accumulation of Cd, Zn, Pb in Taraxacum officinale and Plantago lanceolata from contaminated soils. Proc VII ICOBTE, Uppsala Sv.

Reviewed by J. Bech Borras, Faculty of Biology, Chair of Soil Science, University of Barcelona, Spain.

INDEX

A

B

C

D

E

F

G

N

O

P

Q

R

S

T

U

V

W

X

Y

Z